DEVONIAN FLORAS

Yours sincerely

E. A. Newell Arber

DEVONIAN FLORAS

A STUDY OF THE ORIGIN OF CORMOPHYTA

BY

E. A. NEWELL ARBER, M.A., Sc.D.

TRINITY COLLEGE, CAMBRIDGE
UNIVERSITY DEMONSTRATOR IN PALAEOBOTANY

WITH A PREFACE BY

D. H. SCOTT, M.A., LL.D., Ph.D., F.R.S.

WITH A FRONTISPIECE AND FORTY-SEVEN
FIGURES IN THE TEXT

CAMBRIDGE
AT THE UNIVERSITY PRESS
1921

CAMBRIDGE
UNIVERSITY PRESS

University Printing House, Cambridge CB2 8BS, United Kingdom

Published in the United States of America by Cambridge University Press, New York

Cambridge University Press is part of the University of Cambridge.

It furthers the University's mission by disseminating knowledge in the pursuit of education, learning and research at the highest international levels of excellence.

www.cambridge.org
Information on this title: www.cambridge.org/9781107688452

© Cambridge University Press 1921

First published 1921
First paperback edition 2014

A catalogue record for this publication is available from the British Library

ISBN 978-1-107-68845-2 Paperback

NOTE

THE present critical review of our knowledge of the two Devonian
land floras was one of the last pieces of work which my husband
undertook. Both text and illustrations are embodied in this
book substantially as he left them. His health was already
failing when he finished the manuscript in January, 1918—six
months before his death—but his delight in his subject remained
unabated. I do not think that anything in his scientific life gave
him a keener intellectual pleasure than the development of the
idea—the *Leitmotiv* of the present essay—that the transition
from the Algae to the Vascular Cryptogams no longer remains
a matter of pure conjecture, but that, in the fossil plants of the
Devonian rocks, we witness, actually occurring beneath our eyes,
the passage from the Thallophyta to the Cormophyta. He
welcomed this conclusion as exemplifying a generalisation to
which his experience in research had gradually led him—namely,
that although the apparent insolubility of a problem may be
for many years unhesitatingly ascribed to lack of data, yet, when
the solution is found, it often becomes obvious that the essential
data were all the time under one's hand, and that it was merely
the recognition of their significance that was lacking.

Since the Author left this memoir as a first draft which he
was never able to revise, I must assume the responsibility for
its final form. I am deeply indebted to Dr D. H. Scott, F.R.S.—
who during my husband's life-time was closely associated with
his study of the Devonian floras—for reading the manuscript
and proofs and suggesting a number of emendations, and for

writing the preface and the footnote on p. 45 to which his name is appended. I am also under great obligations to Professor A. C. Seward, F.R.S., for valuable criticism, and to Dr W. T. Gordon for advice on questions relating to Scottish Geology. The portrait which forms the frontispiece of this book is reproduced by kind permission of the Editor of the *Geological Magazine*.

I have to express my gratitude to the Council of the Royal Society for a grant in aid of the preparation and publication of my husband's manuscripts.

AGNES ARBER.

PREFACE

THE present memoir, which I had the advantage of reading in MS. and of fully discussing with the author, seems to me of the utmost interest. A survey of these early Floras of the land is a most useful undertaking and one much wanted at the present time, when important new discoveries have called general attention to the plant-life of the Devonian period. Such a survey is all the more valuable, when, as in this case, there is an underlying theory giving a definite point of view to the exposition of the facts, and animating the whole.

It is a matter for deep regret that the work never received the final revision of the author, and that he never saw the later results of Kidston and Lang's researches on the Rhynie fossils. As it seems to the present writer, the views of those investigators, though differently and more tentatively expressed, are yet in substantial agreement with Arber's, so far as regards the general question of the systematic position of the *Psilophyton* Flora. However this may be, it must be recognised that Arber's conclusions, which were reached altogether independently, bear the stamp of true originality and are absolutely his own.

In the Introduction, the author lays down the essential distinction between the earlier and the later Devonian Floras, the former, called the *Psilophyton* Flora, consisting chiefly of Procormophyta or Propteridophyta, while the latter, the *Archaeopteris* Flora, was chiefly composed of true Pteridophytes. In the first two chapters the geological age and distribution of the two Floras are discussed, and useful tables of genera and species are given.

Chapter III, "Recent Advances in our Knowledge of the *Psilophyton* Flora," gives a clear account of each genus, with the aid of abundant illustrations, collected from various sources. This chapter, and the corresponding one (the fifth) on the *Archaeopteris* Flora, will be of the greatest value to the student,

who will find here, in a small compass, an excellent comparative
description of ancient fossils, often little known, the literature
on which is much scattered and not always readily accessible.
So far as I know, nothing of the kind has been attempted before,
for the Devonian Floras, and I can testify to the great utility
of the author's critical summary.

Attention may be specially called to the evidence, which the
author adduces, in favour of the generic identity of *Psilophyton*
and *Rhynia*, an important conclusion, which will have to be
taken into serious consideration. On the other hand the remarks
on Halle's genus *Sporogonites*, would undoubtedly have been
modified, if the author could have been acquainted with the
later evidence; this point is dealt with in a footnote on p. 45.

Chapter IV, "A Discussion of the Nature and Affinities of
the *Psilophyton* Flora," is of great theoretical importance. The
author expresses his conviction that *Psilophyton* and all the
other genera of that Flora "were much more probably Thallophyta than Pteridophyta" (p. 47). But he also points out that
"*Psilophyton*,...while still Thallophytic in habit, may occupy
anatomically a place half-way between the Thallophyta and
Pteridophyta" (p. 49). This was written early in 1918; it agrees
very nearly with the statement by Kidston and Lang (Part II,
1920, p. 622) that "The facts are consistent with the Rhyniaceae
finding their place near the beginning of a current of change
from an Alga-like type of plant to the type of the simpler
Vascular Cryptogams." Arber had grasped the position at a
time when only a portion of the evidence was before him.

Chapter VI, "The Procormophyta and the Origin of the Cormophyta," completes the exposition of the author's theory. He
holds that there were three distinct main lines of descent among
vascular plants—the Sphenopsida, Pteropsida and Lycopsida,
derived severally from distinct Algal types. The Psilotales he
regards as another entirely separate group, also of Algal origin,
but of geologically very late appearance. This highly polyphy-
letic hypothesis has something in common with the brilliant
speculations of Dr A. H. Church, whose essay on "Thalassiophyta and the Subaerial Transmigration" would have interested
Arber immensely, if he had lived to see it.

We stand at a new point of departure in our theories of the evolution of the higher plants. Arber was one of the first to realise this, and his memoir represents a bold and vigorous effort to grapple with the problems as they presented themselves to him, at the dawn of a new epoch.

It is fair to mention that the last chapter, "On the Origin of the Stele in the Earlier Cormophyta," was regarded by the author himself as quite unfinished; it could hardly have been otherwise with the data at his disposal.

It has been a pleasure to me to write these few lines of appreciation of my old friend's latest work.

D. H. SCOTT.

October 2, 1920.

CONTENTS

LIST OF ILLUSTRATIONS

INTRODUCTION

It is now clear that in Devonian times, two terrestrial floras, quite distinct as regards affinity, existed, one in the earlier part, and one in the later portion of the Devonian period. The former will here be termed the *Psilophyton flora*; it consisted largely, as we hope to show, of Thallophyta belonging, for the most part, to a group now quite extinct which we propose to term the Procormophyta or Propteridophyta. The later flora consisted chiefly, but not entirely, of plants which were obviously Pteridophyta. This assemblage we propose to term the *Archaeopteris flora*.

Our knowledge of both these floras, though still far from complete, has been entirely revolutionised during the last few years by the publication, both at home and abroad, of a series of memoirs, to which we shall presently refer more in detail. These invaluable contributions have necessitated a complete revision of the whole subject, and since no general account of these floras, including these recent advances, at present exists, we propose to commence by a brief enumeration of the characters of their more important genera. We have purposely omitted from our review all the more doubtful types about which little or nothing is known beyond the existence of very obscure or fragmentary examples. We have further in Chapter III of this book confined our attention to a critical summary of the essential features of the morphology and anatomy of the genera belonging to the *Psilophyton* flora, reserving for a separate chapter (p. 46) the entire discussion of the question of their affinities and systematic position. Finally in yet a further Chapter we discuss the very important bearing of these new discoveries on the phylogeny of Cormophyta and in particular of the various Pteridophytic lines of descent.

The researches to which we particularly refer are firstly Kidston and Lang's[1] memoir on the Scottish plant *Rhynia*

---

[1] Kidston and Lang (1917).

Gwynne-Vaughani, the first example of the Psilophyton flora to be known in the petrified state. The importance of this beautifully illustrated memoir can hardly be exaggerated. As we hope to show, it offers us the key to the chief mysteries which have hitherto surrounded the question of the affinities of this type of land vegetation.

Next we have a series of important memoirs by Nathorst[1] and Halle[2] on Lower and Middle Devonian floras from Norway. There is, further, the interesting paper by Don and Hickling[3] on the British Lower Devonian land plant, *Parka decipiens*.

With these recent contributions we may associate others, published earlier, especially the works on Upper Devonian floras by Nathorst[4] from Bear Island and Ellesmereland, by David White[5] from the United States, and last but not least, the very interesting Middle Devonian flora of Bohemia[6].

[1] Nathorst (1913), (1915).   [2] Halle (1916).
[3] Don and Hickling (1917).   [4] Nathorst (1902), (1904).
[5] White (1905), (1907).   [6] Potonié and Bernard (1904).

CHAPTER I

THE GEOLOGICAL AGE OF THE FLORAS

THE question of the Geological age of the floras known from various parts of the world is so vital to the conclusions of this inquiry, that this matter demands special consideration at the outset. So far as we are aware, no one has ever disputed the age of the various beds from which these fossils are derived, with a view to proving them to be younger than Devonian times. Attempts have, it is true, been made in some cases to establish a pre-Devonian age, particularly in Bohemia and Germany. These views are however, we believe, now almost entirely abandoned, and so they need not detain us here. In the great majority of cases, there is stratigraphical or zoological evidence, from the associated sediments, of undoubted Devonian age. Further, as Nathorst and Halle have recently pointed out, the Devonian flora is now so well known, that the age can usually be established by a consideration of the plant remains alone.

Such doubt on the geological side as may exist, relates not to the question of age, but of horizon. This is often a more serious difficulty, but it is not one which is of first importance from the botanical standpoint. We propose, however, to review the evidence as to the horizon, whether Lower, Middle, or Upper Devonian, of our more important Devonian floras, beginning with that of Scotland.

Scotland. The Old Red Sandstone flora of Scotland has now become of particular importance in view of the recent researches on *Rhynia* and *Parka* to which we have drawn attention. The former comes either from the Middle Old Red or from a lower horizon. The latter is only known from the Lower Old Red, so far as Devonian rocks are concerned.

The sub-division of the Scottish Old Red Sandstone has been a matter of some difference of opinion, but the modern view is that the original threefold classification proposed by Murchison in 1859 is correct. Murchison distinguished an Upper, Middle, and Lower division, each characterised by a peculiar fish fauna. This classification is now accepted by the Geological Survey of

Scotland[1] and by those[2] who in recent years have paid particular attention to this question.

In Wales and some parts of Scotland, the Middle Old Red is wanting, and the higher series rests directly, but with marked discordance, on the lower. This, however, is a complication which now presents no geological difficulty.

The correlation of the three divisions of the Old Red with the three sub-divisions of the Devon facies of the Devonian is still, to some extent, uncertain. The key to its solution is to be sought in Russia, where the Middle Devonian includes both the Old Red and Devon facies and faunas. Lower Devonian rocks appear to be absent from Russia, but there is little doubt from the evidence of the Middle Devonian as there developed, that the three horizons in the Old Red correspond at least roughly to the three main divisions of the Devonian[3].

The following correlation, slightly modified from that given by Frech[4], expresses modern views on this point.

Old Red Facies	*Devon Facies*
Upper Old Red (Cheirolepis, Holoptychius and Asterolepis fauna)	Upper Devonian
Middle Old Red (Pterichthys, Coccosteus and Osteolepis fauna)	Middle Devonian
Lower Old Red (Pteraspis, Cephalaspis and Pterygotus fauna)	Lower Devonian

---

[1] Hinxman and Grant Wilson (1902) (see especially Appendix, Part I, Palaeontological, pp. 81–83, by R. H. Traquair); Crampton and Carruthers (1914); Horne and Hinxman (1914).

[2] Hickling (1908), p. 396; Macnair and Reid (1896).

[3] Hickling (1908) and the references there quoted.

[4] Frech (1897), p. 123.

The horizon of the plant-bearing beds in the Old Red of
Scotland can in most cases be determined by the associated
fish faunas. Important specimens, including *Psilophyton* in a
petrified condition[1], recently discovered in the Dryden Shales at
Rhynie, Aberdeenshire, by Dr Mackie[2] are, however, exceptional
in that the precise horizon has not yet been ascertained. These
beds are, however, regarded as not younger than Middle Old
Red[3], but since this genus is known to range throughout
Devonian time, the precise horizon of these beds is immaterial
from this point of view.

Ireland. The plant-bearing sandstones of the Old Red of the
South of Ireland are referred to the Upper Old Red (Upper
Devonian) on the evidence of the associated fish remains.

England. The few plants known from the Devonian of England
come from the type beds of the Upper Devonian (Baggy or
Cucullaea beds) in North Devon.

Belgium. Two plant-bearing horizons occur in Belgium which
on stratigraphical grounds are assigned to the Lower and Upper
Devonian respectively. These are the "Poudingue de Burnot"
and the "Psammites du Condroz."

Germany. A small but important flora from near Herborn
in Hesse-Nassau has been recently referred to the Silurian. On
the fossil plant evidence, however, there can be no doubt that
it is of Devonian age, and belongs, in all probability, to the
Upper Devonian.

Bohemia. The large flora from the horizon (h 1) of Barrande's
system of classification of the Devonian rocks of Bohemia was
likewise originally referred to the Silurian. The Devonian age of
these beds is now, however, admitted. More recently it has been
found[4] that *Stringocephalus Burtini*, a characteristic Middle
Devonian fossil, occurs on a yet higher horizon (h 3) and thus
the plant-bearing beds clearly also belong to this same zone.

Norway. The evidence as to the horizon of the plant-bearing
beds in Eastern and Western Norway is purely palaeobotanical.
There is no zoological evidence. Nathorst, as we think rightly,

---

[1] For the case in support of this identification, see pp. 24–26.
[2] Mackie (1914); Horne and Mackie (1917).
[3] Kidston and Lang (1917), p. 762, footnote.   [4] Jahn (1903).

referred the Western Norwegian flora to the Middle Devonian, while Halle assigned that of Röragen to the Lower Devonian.

Russia. In the Devonian of the Donetz Basin in Russia, plants occur associated with a fauna believed to be Upper Devonian.

Bear Island and Ellesmereland. The Upper Devonian horizon of these beds is determined by the occurrence of fish remains.

Spitzbergen. In Spitzbergen two floras on different horizons occur associated with fish remains. That of Mimers-Thal is probably Upper Devonian, that of Dickson Bay, Lower Devonian.

Canada. In New Brunswick, plants occur on at least two horizons, but until these beds have been more carefully studied, it will not be certain whether the higher plant-bearing beds are of Middle or Upper Devonian age. The lower plant horizon of Gaspé is, however, undoubtedly Lower Devonian, as the associated fish remains clearly indicate.

United States. Plant-bearing beds occur on several horizons in the United States. Of these, the best known at present is the Upper Devonian flora of the Perry Basin (S.E. Maine) probably referable to the Chemung series. In New York State and elsewhere other plants occur, some of which may be older than the Upper Devonian.

Australia. In New South Wales and Victoria, Devonian plant-bearing rocks occur, some of which are believed on stratigraphical grounds to be of Upper Devonian age. Certain plants may even occur on lower horizons, but naturally a strict correlation between these far distant rocks and the Devonian horizons of Europe is a difficult matter.

The above are here relied upon as the best known Devonian floras of which the horizons have been more or less satisfactorily determined. Isolated members of these floras, particularly examples of *Psilophyton*, occur in many other countries, such as France and China, but these are not taken into consideration here.

*Summary of the horizons of the chief Devonian
Plant-bearing beds.*

Upper Devonian:
Scotland (Upper Old Red)
Ireland   „   „   „
England (Marwood beds)
Belgium (Psammites du Condroz)
Germany (Hesse-Nassau)
Russia
Bear Island
Ellesmereland
Spitzbergen
United States
? Canada
Australia

Middle Devonian:
Scotland (Middle Old Red)
Western Norway
Bohemia (Étage *h* 1)

Lower Devonian:
Scotland (Lower Old Red)
Belgium (Poudingue de Burnot)
Norway (Röragen)
Spitzbergen
Canada (Gaspé)

CHAPTER II

THE TWO DEVONIAN FLORAS

As we have already stated, it is clear that two distinct floras existed during Devonian times, not side by side, but successively. This is clear from a comparison of the dominant genera of the three horizons in that series. Among the earlier types, such genera as *Psilophyton*, *Arthrostigma* and *Hostimella*[1] are prominent and many of the plants of the Upper Devonian period are entirely wanting. This we propose to term the *Psilophyton flora*. It was a flora not by any means sharply marked off from that which preceded, or that which succeeded it. In the earlier Devonian rocks, we find not only the dominant members of this type of flora, but also survivals of a still earlier flora of which *Cryptoxylon*, *Nematophycus* and *Pachytheca* are examples. We do not say that these types are ever found in the same beds as *Psilophyton*—all we remark is, that they existed at the same period.

The land flora of Silurian times is at present almost unknown, but we are acquainted, in *Parka*, with at least one British genus of the Psilophyton flora, which goes back as far as the Silurian. It has been also stated by Dawson that *Psilophyton* itself occurs in the Upper Silurian of Canada. It would thus seem that the flora of the close of Silurian times, whether marine or terrestrial, had much in common with that of the earliest stage in Devonian history. We know of at least four genera common to these two formations.

As we pass upwards from the lowest sediments of Devonian age, we find that the members of the Psilophyton flora begin to die out and that their place is taken by new arrivals, which, as

---

[1] [The spelling *Hostimella* and not *Hostinella* is used here, since Jahn (1903), p. 74, shows that the former is correct. See also Potonié, H. and Bernard, C. (1904). p. 11. A. A.]

we shall show here, are of an entirely different morphological nature. By the time we reach the Upper Devonian, these newer types, the *Archaeopteris flora*, have become dominant in their turn. *Archaeopteris, Sphenophyllum, Bothrodendron*, among many other genera, are all clearly Pteridophyta. The members of the Psilophyton flora were Thallophyta, as we hope to show here.

In Upper Devonian times, some members of the *Psilophyton flora* were still in existence, though in greatly reduced number and in a position of subordination to the dominant Archaeopteris flora. By the time we reach the next higher series, the Lower Carboniferous rocks, the Psilophyton facies has entirely disappeared. On the other hand, during this period, the Archaeopteris facies reaches its maximum development, and it persisted unchallenged as regards dominance until the close of this epoch, while many survivals lingered on even into Coal Measure times. We have then, in the Lower Devonian, a very ancient land flora, which in Middle and Upper Devonian times was gradually displaced by a new flora which only reached its maximum development in the earlier part of the Carboniferous epoch. Thus, while the flora of the Upper Devonian is essentially of the Carboniferous facies, that of the Lower Devonian is of a quite different archaic type. The following table shows the distribution in Devonian time of the two floras.

A Summary of the Chief Genera of the Archaeopteris and Psilophyton Floras with their distribution in Devonian Time.

	Archaeopteris Flora		Psilophyton Flora
UPPER DEVONIAN	*Sphenophyllum* *Pseudobornia*	} Sphenopsida	*Psilophyton* *Ptilophyton*
	Psygmophyllum	Palaeophyllales	*Thursophyton*
	Archaeopteris *Rhacopteris* *Sphenopteris* *Sphenopteridium* *Cephalopteris* ? *Cordaites*	} Pteropsida	*Barrandeina* *Barinophyton* *Taeniocrada*
	Bothrodendron *Archaeosigillaria* *Leptophloeum*	} Lycopsida	

Archaeopteris Flora	Psilophyton Flora

MIDDLE DEVONIAN

Archaeopteris Flora	Psilophyton Flora
Hyenia, Sphenopsida	*Psilophyton*
Psygmophyllum, Palaeophyllales	*Bröggeria*
	Thursophyton
	Pseudosporochnus
Sphenopteridium, Pteropsida	*Hostimella*
	Barrandeina
Archaeosigillaria ⎫	
Leptophloeum ⎬ Lycopsida	
Protolepidodendron ⎭	

LOWER DEVONIAN

Archaeopteris Flora	Psilophyton Flora
? *Leptophloeum*, Lycopsida	*Psilophyton*
	Arthrostigma
	Parka
	Hostimella
	(*Cryptoxylon*)

We now propose to review the distribution of the above genera in the standard Devonian floras indicated in Chapter I.

SCOTLAND.

A revised Summary of the Devonian Flora of Scotland[1] occurring in the Old Red Sandstone with the dates of their first record from Scotland.

UPPER OLD RED SANDSTONE (Caithness)[2]:

> *Ptilophyton Thomsoni*, Dawson[3], 1878 (includ. *Caulopteris Peachii*, Salter[4], 1859).
> *Thursophyton Milleri*, (Salter)[5], 1858.
> *T. Reidi*, (Penhallow)[6], 1892.
> *Archaeopteris hibernica*, (Forbes)[7], 1857.

MIDDLE OLD RED SANDSTONE (Cromarty)[8]:

> *Psilophyton princeps*, Dawson[9], 1841.

---

[1] This list chiefly differs from that given by Kidston in 1902 (Kidston (1902)) in regard to the horizons to which the fossils are ascribed, which we have given in accordance with the conclusions of the Scottish Geological Survey.   [2] Crampton and Carruthers (1914).

[3] Carruthers (1873), Pl. 137; Dawson (1878), p. 385 and Pl. IV.

[4] Salter (1859), p. 408, Fig. 14 a.

[5] Salter (1858), p. 75; Penhallow (1892), p. 5; Nathorst (1915), p. 17.

[6] Penhallow (1892), p. 8, Pl. I, fig. 2; Reid and Macnair (1899), Pl. XXII; Nathorst (1915), p. 19.

[7] Miller (1857), p. 454, Fig. 124.   [8] Horne and Hinxman (1914).

[9] Miller (1841), p. 100, Pl. VII, figs. 3–5; Miller (1857), Fig. 119, p. 429, Fig. 120, p. 432; Dawson (1859), p. 481.

LOWER OLD RED SANDSTONE (Forfarshire, Perthshire, etc.)[1]:
Psilophyton princeps, Dawson[2], 1859.
P. ornatum, Dawson[3], 1871.
P. robustius, Dawson[2], 1859.
Arthrostigma gracile, Dawson[4], 1871.
Zosterophyllum myretonianum, Penhallow[5], 1892.
Parka decipiens, Fleming[6], 1831.
Cryptoxylon Forfarense, Kidston[7], 1897.

The great peculiarity of the Old Red Sandstone flora of
Scotland, so far as it is at present known, is its wealth of members
of the Psilophyton flora and its poverty in examples of the
Archaeopteris flora in its highest stage.

Ireland. The Upper Devonian flora of the South of Ireland
includes several species of *Archaeopteris* (*A. hibernica*, Forbes[8,11]
and *A.Tschermaki*, Stur[9]), *Bothrodendron Kiltorkense*, (Haught.[10]),
and *Sphenopteris Hookeri*[11], (Bailey). No members of the older
archaic flora have yet been recognised.

England. The Upper Devonian of Devonshire[12] has yielded
Sphenopteridium rigidum, a fragmentary *Sphenopteris*, a unique
fructification, *Xenotheca*, a *Telangium* and a doubtful leaf of
Cordaites; there are no traces of archaic forms.

Belgium. The Upper Devonian rocks of Belgium[13] contain
Archaeopteris, *Sphenopteridium condrusorum*, and a species of
Sphenopteris, associated with *Barinophyton*, and possibly
Psilophyton. The Lower Devonian flora[13] of the same country is
very obscure. *Arthrostigma* (the so-called *Lepidodendron Gas-
pianum*) and *Psilophyton* probably occur.

Germany. In the Upper Devonian of Hesse-Nassau[14], *Sphenopteridium rigidum*, an *Archaeopteris* (the *Sphenopteris densepin-
nata* of Ludwig) and a *Rhacopteris* (the *Cyclopteris furcillata*,
etc., of Ludwig and *Triphyllopteris* of Schimper) are possibly

[1] Hickling (1908).   [2] Dawson (1859).
[3] Included hitherto as a variety of Dawson's *P. princeps*.
[4] Dawson (1871), p. 41; Kidston (1894), p. 102.
[5] Penhallow (1892), p. 9, Pls. I, II.
[6] Fleming (1831); Don and Hickling (1917).
[7] Kidston (1897).
[8] Carruthers (1872); Kidston (1888) and (1906).
[9] Johnson (1911[2]).   [10] Johnson (1913).
[11] Kidston (1906).   [12] Arber and Goode (1915).
[13] Crépin (1874) and (1875); Gilkinet (1875[1]), (1875[2]).
[14] Ludwig (1869); Potonié (1901).

associated with *Psilophyton* (? the *Palaeophycus gracilis* or *Noeggerathia bifurca* of Ludwig).

Bohemia. The Middle Devonian flora of Barrande's horizon *h* 1 in Bohemia[1] consists almost entirely of members of the Psilophyton flora, including *Arthrostigma*, *Hostimella* and probably *Psilophyton*, with *Pseudosporochnus*, *Barrandeina*, *Protolepidodendron* and *Thursophyton*. The only more modern types are a doubtful example of *Sphenopteridium* and *Archaeo-sigillaria* (the *Protolepidodendron Scharyanum* of Krejči and Potonié).

Norway (Röragen). The probably Lower Devonian flora of Röragen[2] comprises *Arthrostigma*, *Psilophyton*, *Hostimella*, *Aphyllopteris*, and *Sporogonites*, without any more modern types.

Norway (Western). The possibly Middle Devonian of Western Norway[3] includes *Spiropteris*, *Aphyllopteris*, *Thursophyton*, *Bröggeria*, and possibly *Barrandeina*, associated with the more modern types, *Psygmophyllum* and *Hyenia*.

Russia. From the Upper Devonian of the Donetz Basin[4] are known two species of *Archaeopteris*, examples of an isolated fructification (*Dimeripteris*) with a species of *Sphenopteris* and possibly a *Lepidodendron*. No members of the Psilophyton flora occur.

Bear Island (Arctic Regions). We find here a very interesting Upper Devonian flora of the Archaeopteris type, including *Archaeopteris Roemeriana*, (Goepp.) and other species, *Bothro-dendron Kiltorkense*, (Haught.), species of *Sphenopteridium*, *Sphenophyllum* and *Stigmaria* associated with the rare types *Pseudobornia ursina*, Nath., and *Cephalopteris mirabilis*, Nath.[5]. No archaic types have been recognised.

Ellesmereland (Arctic Regions). From the Upper Devonian of Ellesmereland[6] two species of *Archaeopteris* (*A. Archaetypus* and *A. fissilis* which both occur in Russia) are associated with a *Sphenopteridium*.

Spitzbergen (Arctic Regions). In rocks assigned to the Upper (or possibly the Middle) Devonian of Spitzbergen[7], we find a

---

[1] Potonié and Bernard (1904); Stur (1881).
[2] Halle (1916); Nathorst (1913).   [3] Nathorst (1915).
[4] Schmalhausen (1894).   [5] Nathorst (1902).
[6] Nathorst (1904).   [7] Nathorst (1894).

Psygmophyllum and a *Leptophloeum* and perhaps a *Bothrodendron*. From Lower Devonian beds, a *Psilophyton*-like plant is known, associated also with a doubtful *Psygmophyllum* (the *Cyclopteris* sp. of Nathorst).

United States and Canada. In the Perry Basin[1], South-east Maine, we find *Psygmophyllum Brownianum*, (Dawson), several species of *Archaeopteris* (*A. Jacksoni*, Daws., *A. Rogersi*, Daws., etc.), species of *Sphenopteridium* and *Dimeripteris*, associated with archaic types such as *Psilophyton, Barrandeina, Barino-phyton, Leptophloeum*, etc. This flora is Upper Devonian, pro-bably Chemung, in age. In New York State, fossil plants (*Archaeosigillaria*) occur in the Portage group of the Upper Devonian[2], and in the Middle Devonian of Ohio[3] (Corniferous Limestone) where *Barrandeina* (the *Caulopteris* spp. of Newberry) is associated with a *Leptophloeum* and petrified wood. Other examples are also known, but are neglected here, since the records are somewhat obscure. In Canada, a very similar flora to that of Perry occurs in the Upper Devonian[4]; but this is in urgent need of revision. With regard to the Lower Devonian flora of Gaspé, it is undoubted that species of *Psilophyton* and *Arthrostigma*, and probably other types occur associated with *Nematophycus*. But until this flora also has been revised, the list of genera occurring on this horizon must remain somewhat uncertain.

Australia. In Devonian rocks in Australia, especially in Victoria and New South Wales, *Leptophloeum australe* (usually known as *Lepidodendron australe*) is frequent in beds assigned with certainty to the Upper Devonian[5], and possibly also in others of Middle Devonian age. On the Genoa River, Auckland[6], a fragmentary leaf like that of *Cordaites* occurs, associated with *Archaeopteris*, and *Sphenopteris*. The only archaic type known is *Barinophyton*. A similar flora of *Archaeopteris* (*A. Howitti* and *A. Wilkinsoni*), *Sphenopteris*, and *Cordaites* is associated with *Leptophloeum* and possibly *Bothrodendron* in Victoria, but without any archaic forms so far as is known.

---

[1] White (1905).   [2] White (1907).   [3] Newberry (1889).
[4] Dawson (1859) and (1871), excluding the fossil plants from St John's, N.B., which are of Upper Carboniferous age.
[5] David and Pittman (1893).   [6] Dun (1897).

CHAPTER III

RECENT ADVANCES IN OUR KNOWLEDGE OF THE MORPHOLOGY AND ANATOMY OF MEMBERS OF THE PSILOPHYTON FLORA

It has been known since the days of Hugh Miller that in the Old Red Sandstone of Scotland a number of simple if somewhat obscure plant remains are to be found. A similar flora was first described by the late Sir William Dawson, from the Lower Devonian of Canada in 1859. Halle[1] has recently pointed out that these are "the remains of the very oldest land-flora at present known; and it may be stated at once that there is a far greater difference between this flora and that of the Upper Devonian than between the latter and the Lower Carboniferous."

These fossils, occurring as impressions with only slight traces of their original anatomical structure, have until recently been generally regarded as very doubtful objects and much scepticism has been expressed by botanists as to the morphological interpretation of the Scottish and Canadian plants given by Dawson, Penhallow and others. As we now know, this scepticism has been largely misplaced. Except in the matter of affinities, on which point the evidence hitherto has always been very slender, the earlier accounts of these fossils were extremely accurate. The fact that these plant remains are apparently of a simple type of habit has often been explained by an appeal to the imperfection of the record. Such fossils were commonly regarded as mere petioles or rachises of fronds which had been so damaged before preservation that no trace of the lamina now exists.

It is now, however, quite clear that this conception is fundamentally erroneous. We know now, in several cases, what practically the whole of the plant body, and it is clear that instead of dealing with fragments of Cormophytes, as was formerly supposed, we are in reality confronted with a vegetation occupying a lower place in the scale of plant evolution.

In order to make these matters clear, we propose in the present

[1] Halle (1916), p. 4.

chapter of this memoir to pass in review the main facts relating to the morphology and anatomy of these fossils. In each case we begin with a summary of what appear to us to be the critical features of each genus and we then pass on to a discussion, also of a critical nature, of the more recent advances in our knowledge of each type.

It should, however, be clearly understood that while some of the genera here discussed are now placed on a firm scientific footing, many others remain extremely obscure, and are as yet only of minor interest. In the matter of the literature, we have contented ourselves as a rule by quoting only the latest of a series of memoirs dealing with each subject. References to the earlier literature will be found in the papers quoted.

PSILOPHYTON (including RHYNIA and DAWSONITES).

(Figs. 1–7.)

Psilophyton, Dawson, 1859. Terrestrial plants, consisting of a rhizome from which dichotomously branched, erect axes arise, the terminations of which are circinately coiled in the young state. Shoots leafless, vascular, possessing stomata, and emergences, the latter being either macro- or microscopic. Fructification consisting of sporangia borne terminally on some of the erect shoots; wall of sporangium multi-layered.

Fig. 1. *Psilophyton*. Dawson's restoration of the sterile thallus, published in 1859. This restoration is correct except in the matter of the lateral organs borne on the longest (central) axis. After Dawson (1859).

Distribution. Devonian and perhaps Silurian. In Scotland *Psilophyton* apparently occurs only in the Lower and Middle Old Red. It however occurs in Upper Devonian rocks in the United States and possibly in Belgium and Germany. It is thus clear that the genus is distributed throughout Devonian times. According to Dawson[1] it also occurs in Silurian rocks of Canada.

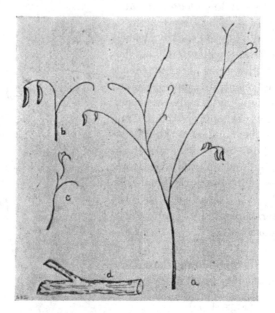

Fig. 2. *Psilophyton.* Fertile fragments restored by Reid and Macnair in 1896. *a, b,* fertile axis; *c,* circinate vernation; *d,* rhizomata. Considerably reduced. After Reid and Macnair (1896).

Several species of *Psilophyton* variously regarded as quite, or as probably, distinct, occurring in Scotland and Canada, were first discriminated by Dawson in 1859 or in 1871. Unfortunately on the subject of species of this genus there is at present considerable confusion and a critical revision of *Psilophyton* is now, in the light of *Rhynia,* more than ever needed. The difficulty here lies in the fact that it can only be attempted by some

[1] Dawson (1871).

authority resident in America where the type specimens of Dawson are located.

According to modern opinion, Dawson[1] in 1871 included at least two species under the term *P. princeps*. Confusion has also arisen from the attribution by Carruthers[2] in 1873 of some of the Scottish examples to Goeppert's *Haliserites Dechianus*, as

Fig. 3. *Psilophyton princeps*, Dawson. Type specimens of Dawson's variety "*ornatum*," with macroscopic emergences. The left-hand figure shows the circinate vernation. After Dawson (1871).

P. Dechianum (Goepp.). In fact at the present time the Scottish plant is probably better known under that name than by any other. These difficulties may be overcome as follows. Carruthers' determination should be completely ignored. In our opinion it is erroneous. Further the specimens which he figures do not belong to *Psilophyton* at all, but to a distinct genus *Ptilophyton*.

[1] Dawson (1871).   [2] Carruthers (1873).

The difficulty in regard to Dawson's species will, we believe, vanish, if it can be shown, as we shall attempt to demonstrate

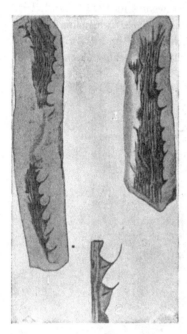

here, that the presence or absence of *macroscopic* emergences or so called spines is a matter of no systematic importance. The species *Psilophyton princeps* should include those erect shoots with fairly stout spine-like emergences (Fig. 3), those on which the emergences are small delicate structures (Figs. 4 and 5) and, further, those stems which are apparently smooth and without macroscopic emergences of any kind. This it may be remarked is exactly the view of the importance of these structures which Dawson himself urged.

He states[1] explicitly that most observers would separate specifically the two types *P. princeps* and *P. ornatum*, but he believes they pass into one another and cannot be clearly separated on these grounds. We may therefore enumerate the species of impressions of *Psilophyton* occurring in Scotland as follows:

Fig. 4. *Psilophyton princeps*, Dawson, from the Lower Devonian of Röragen, Norway. 1. Attributed by Halle to the genus *Arthrostigma*, but described as a "narrow, *Psilophyton*-like stem." 2. Part of "1" twice enlarged to show nerves of emergences. 3. Attributed by Halle to *Psilophyton princeps*, or possibly to *Arthrostigma*. After Halle (1916).

(1) *P. princeps*, Daws.[2] (including the variety *P. ornatum*, Daws.[3]) (Figs. 3–5). Erect shoots, slender or of medium thickness, dichotomously or laterally branched, forks wide, bark macroscopically smooth or covered with scattered, or numerous

---

[1] Dawson (1871), p. 39.   [2] Dawson (1871), Pl. IX, figs. 102–108.
[3] Dawson (1871), Pl. IX, figs. 97–101, 104, 104 a, 109–110; Pl. X, figs. 112–114, 118.

and crowded, small chaffy scales or larger spine-like processes. Tips of branches when young circinately coiled. Erect shoots proceeding from a horizontal rhizome bearing rhizoids. Shoots vascular, bearing stomata. Sporangial wall multi-layered, sporangia borne on the finer terminations of some of the younger macroscopically smooth shoots, singly or in pairs (Fig. 6, p. 21).

(2) *P. robustius*, Daws.[1]. Stems rather stout, bark smooth or slightly furrowed without macroscopic emergences, branching chiefly lateral or pseudo-dichotomous when terminal. Terminations of branches bearing sporangia in clusters.

Fig. 5. *Psilophyton princeps*, Dawson. Type specimens (1859) with small scale-like emergences, two of the specimens showing the circinate vernation. After Dawson (1859).

(3) *P. elegans*, Daws.[2]. Axes very slender, dichotomously branched, produced in tufts from thin rhizomes. Surface smooth, with very delicate wrinkles, but without macroscopic emergences. Fructification (?) believed to consist of small oval bodies borne below the bifurcations of the axes.

More recently important observations have been published on *Psilophyton* by Solms Laubach[3], and David White[4]. The former recognises *P. princeps* alone as a good species. Dawson's other types of the same genus are regarded as indefinable. White[4] discriminates between a spiny type of *Psilophyton* (*P. ornatum*)

[1] Dawson (1871), Pl. XII.
[2] Dawson (1871), p. 40, Pl. X, figs. 122, 123.
[3] Solms Laubach (1895), p. 76.
[4] White (1905), p. 61.

and a smooth type with characteristic costation and no evidence of spines or scales. The latter is a more lax type, and freely branched.

The most recent work bearing on this genus is that of Halle on Lower Devonian specimens from Norway, and of Kidston and Lang on the Scottish Devonian *Rhynia*. We will now consider these very important contributions in some detail beginning with Halle's conclusions.

Halle[1], in an elaborate attempt to apply some definite meaning to the term *Psilophyton*—an attempt with which at the time it was written we were much in sympathy—would wish to confine this term to those stems alone which, in whole or part, bear spine or leaf-like organs. He says "in order to establish an acceptable definition of the genus *Psilophyton*, it is necessary to confine its use to stem-like structures bearing spines or small leaves. Isolated branch-systems without spines...cannot be regarded as belonging to *Psilophyton* unless they are found in actual connection with spine-bearing *Psilophyton*-stems[2]." The non-spinous stems bearing fructifications (Fig. 6), which Dawson referred to *P. princeps*, are removed by Halle to a distinct genus *Dawsonites*, as *D. arcuatus* n. spec.; according to Halle the term *P. princeps* should be used only for the spiny type of stem (Fig. 3), the fructification of which he asserts is not as yet known.

Throughout Halle's criticisms it is clear that he shared in no small degree the doubt which others had long cast on the correctness of Dawson's morphological and taxonomic conclusions. It is perfectly true of course that Dawson did not prove, by means of incontestable figures, many of his statements, in the manner which we have learnt to expect in modern research. It has also to be borne in mind that when Halle wrote his memoir, he did not know of the entirely new light which *Rhynia* has since shed on these questions.

It must however be confessed that even at that time there were no just grounds for discriminating species merely on the presence or absence of macroscopic scale or spine-like emergences. Several species of *Psilophyton* (*P. robustius*, *P. elegans*, etc.) were already known in which no such emergences are found, and thus

---

[1] Halle (1916).   [2] *Ibid.* p. 22.

the removal of the smooth axes bearing fructifications to a separate genus could hardly be justified. Halle[1] himself describes a new species (*P. Goldschmidtii*) in which the axes below were spinous, though without visible macroscopic emergences in the higher

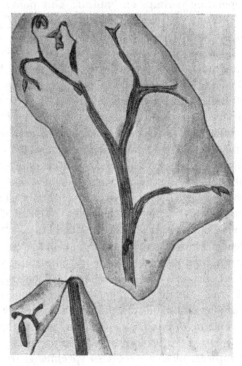

Fig. 6. *Psilophyton princeps*, Dawson, from the Lower Devonian of Röragen, Norway. Fertile axes (the *Dawsonites arcuatus* of Halle) without macroscopic emergences. After Halle (1916).

parts. We shall return to this point a little later when we shall endeavour to show that in *P. princeps*, the axes always bear emergences, though sometimes they are of microscopic size. We may agree with Halle that the spiny shoots (*P. ornatum* of Dawson) are not known in the fertile state. These shoots are

[1] Halle (1916), p. 21.

probably old shoots and the old shoots may have been sterile. At any rate we think it probable that the degree to which the emergences are *visible* depends partly on the age of the shoots, though it may, as Dawson states, be also a very variable character, depending perhaps on habitat. Our point, however, is that a fossil is not justly excluded from the genus *Psilophyton* merely on the ground that it has no *visible* emergences.

We may also agree that Halle[1] was the first to discover the vascular nature of the spiny type of stem (*P. ornatum*), but it was shown by Dawson and has been more recently confirmed by Kidston and Lang that the *apparently* spineless stem is also vascular. The fructifications of what we here term *P. princeps* (figured by Halle[2] under the name *Dawsonites arcuatus*, sp. et gen. nov.) are the best examples we know in the form of impressions (Fig. 6). They are described as "terminal capsules of a narrowly obovoid or short fusiform shape and usually 3–5 mm. long." Spores have not been recognised in them, a fact which has become immaterial in view of the fuller evidence of the same organs which we now possess in the petrified state (*Rhynia*).

We now reach the most recent contribution to the subject of *Psilophyton*, and undoubtedly the most important yet made, namely a recent account of a member of this genus published by Kidston and Lang[3] under the name *Rhynia Gwynne-Vaughani*. For reasons which will be fully discussed a little later, we have no hesitation in referring *Rhynia* to *Psilophyton* and this species is in all probability either *P. princeps* or *P. elegans* as here defined.

The great interest of these Scottish specimens, from a chert bed, not younger than the Middle Old Red, at Rhynie in Aberdeenshire, is that the plants are not only petrified but complete. They occur in a most remarkable series of beds of silicified peat, crowded with stems of this plant *in situ*. The description of the habit and morphology of *Rhynia* given by Kidston and Lang confirms in a remarkable manner the account

---

[1] Halle (1916). The xylem elements are described by Halle, as by Dawson, as scalariform, whereas in *Rhynia* they are annular, but from impressions and macerated material it must be very difficult to distinguish between these two types of thickening.

[2] Halle (1916), p. 25, Pl. 3, figs. 1–9, Pl. 4, figs. 18–21.

[3] Kidston and Lang (1917).

of *Psilophyton* given by Dawson as far back as 1859, on which so much scepticism has since been expressed. There is the same habit (Figs. 1, 2, pp. 15, 16)—branched underground rhizomes attached to the peaty soil by numerous rhizoids, and bearing erect aerial shoots, eight inches or more in length and ranging from 6 mm. to under 1 mm. in diameter. These shoots bore no leaves. They occasionally branched dichotomously. They bore "small hemispherical projections which were more or less closely placed without apparent regularity. On some of these bulges tufts of rhizoid-like hairs were borne, while in other cases the projections developed into adventitious branches,...some aerial axes ended in large, elongate-pointed sporangia[1]."

The most interesting matter in connection with these stems is that the anatomy, which is very perfectly preserved in silica, is often but not always vascular. Here again we have an important confirmation of a fact which Dawson as early as 1859 had been able to ascertain from simple impressions of similar stems. There is a small central strand of xylem (annular tracheids), surrounded by phloem and externally by a wide parenchymatous cortex and a well-marked epidermis. There is no distinct endodermis or pericycle. Stomata[2] occur on the aerial stems, but they do not appear to have been frequent. Some slender axes are also met with in which no vascular tissues are developed. Sieve plates have not been recognised in the phloem region. All the tracheids of the xylem are alike in size, there being no obvious protoxylem groups and no conjunctive parenchyma.

The branches were vascular, but apparently their steles were not connected with that of the main axis.

The sporangia varied considerably in size, but seem to have attained a length of at least 12 mm. Their walls were several layers thick. They contained an enormous number of spores.

As Dawson originally pointed out, these plants were *land plants*, and Kidston and Lang's[3] account of the habitat of *Rhynia* as "a peaty soil practically composed of the decaying remains of the same species" agrees fundamentally with

[1] Kidston and Lang (1917), pp. 765–6.
[2] Dawson (1871), p. 90, stated that stomata occurred but he did not figure them clearly on Pl. XI, fig. 129.
[3] Kidston and Lang (1917), p. 774.

Dawson's earlier conclusions on this point. These authors also all agree in finding in *Psilotum* the nearest existing type of habit.

We now pass on to consider the question of the identity of *Rhynia* with *Psilophyton*. These authors, after discussing this question, conclude that the two genera are distinct. They compare *Rhynia* with *P. princeps* and conclude that the latter differs from the former "in the presence of spines, in the more profuse dichotomous branching, in the subordination of some of the branches to a sympodial main axis, and in the absence, so far as we know, of lateral adventitious branches[1]."

The weak point of this argument is that it does not take into account species of *Psilophyton* other than *P. princeps*. As we have seen, several other species of this genus, e.g. *P. elegans* and *P. robustius*, have neither visible spines nor scales, while the latter has "slender alternate branches" arising from a relatively robust axis[2]. The agreement of *Rhynia* with Dawson's account of *P. princeps* and *P. elegans* appears to us to be so close that we have no doubt as to a generic identity, at least, existing between these types.

In support of this contention as to the generic identity of *Rhynia* with *Psilophyton*, we have one constructive addition to make to the discussion, and this is perhaps important. *Rhynia* was not a spineless type, despite Kidston and Lang's assertion. The small "lateral protuberances or bulges" of those authors *are the spine-like emergences*. In order to realise this, it is only necessary to compare the Figs. 7 and 8 on Pl. III of the paper of these authors, which show the surface of stems of *Rhynia* enlarged fourteen times, with the similar figures of Halle's *Psilophyton princeps* (i.e. the very spiny type, Dawson's *P. ornatum*) on Pl. 2, figs. 3 and 5 of Halle's paper, which are of nearly equal magnification (× 16). These figures are reproduced here in Fig. 7. This comparison is we think conclusive, and it also settles once and for all the correlation of the external morphology of the spines with their internal anatomy.

We may add that on microscopic examination, certain impres-

[1] Kidston and Lang (1917), p. 779. [In a more recent paper, *Trans. Roy. Soc. Edinb.* Vol. 52, 1920, p. 603, Kidston and Lang have described a second species of *Rhynia, R. major* which shows no trace of adventitious branches. A. A.]
[2] Dawson (1859), Fig. 2 a, p. 481.

sions of *P. princeps*[1] (the nearly smooth type) from the Middle
Old Red of Cromarty in Scotland, collected by the present author
some years ago, appear to show indistinct indications of the
emergences comparable to those of *Rhynia* and Halle's specimens.

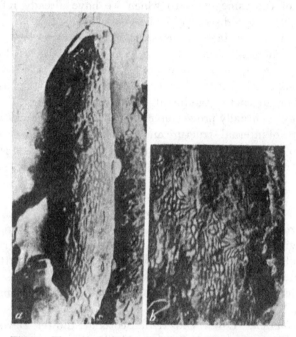

Fig. 7. The proof of the identity of *Psilophyton* and
Rhynia. (*a*) Enlarged surface view of a petrified axis of
Rhynia Gwynne-Vaughani showing surface sculpture and
several emergences ("the small projections or bulges" of
Kidston and Lang) (× 14). After Kidston and Lang
(1917). (*b*) Enlarged surface view of an impression of
Psilophyton princeps showing surface sculpture and two
emergences (the "spines" of Halle) (× 16). After Halle
(1916).

It is thus we think clear that all known examples of *P.
princeps* bore scales or spines, and the anatomy of these structures
shows clearly that they are morphologically emergences and
neither leaves nor branches. In many cases, however, especially
near the apex of the shoots, these emergences are so small and

[1] No. 25 Devonian Plant Coll., Sedgwick Museum, Cambridge.

scattered that they are invisible macroscopically. As they increase in size and in number, they gradually become more and more macroscopic, at first resembling chaffy scales, later spinous outgrowths. These facts explain the different appearance of shoots of the same plant to which we have already referred (pp. 21, 22) when discussing Halle's conclusions.

In view of these facts, impressions of *Psilophyton* which, when without emergences, must appear to be very featureless fossils, can no longer be regarded as doubtful objects or as unworthy of serious consideration. The genus is now as important as any from a phylogenetic standpoint, as we shall see later.

It may eventually prove convenient to retain the term *Rhynia* as a type of internal structure and if this is necessary such would seem to be the chief justification of its existence. As a generic term *Psilophyton* has undoubted priority.

ARTHROSTIGMA.

(Figs. 8, 9.)

Arthrostigma, Dawson, 1871[1]. Axis very stout, bifurcating and giving off lateral members, irregularly furrowed or ribbed longitudinally, bearing numerous large and long scattered, straight, sometimes falcate, spine-like organs. Axes possessing a slender central strand of vascular tissue. Fructification unknown.

Distribution. Lower Old Red, Scotland; Lower Devonian, Canada, Norway, and (?) Belgium; Middle Devonian, Bohemia.

According to Halle[2], this plant displays considerable variations in the shape of the spine-like organs and their manner of attachment. There are first of all cylindrical impressions with a radial arrangement of their leaf-like projections, though, in a few instances, cases of a pseudo-verticillate arrangement occur. "There can sometimes be noted a very fine but distinct vein-like line running through the leaf,"...which "no doubt represents a vein or vascular strand." No leaf scars on the stem can be seen in the case of the Norwegian specimens.

Secondly, there are stems with "unusually densely and regularly placed leaves." A third type consists of stouter specimens "with

---

[1] Dawson (1871), p. 41.   [2] Halle (1916), p. 7.

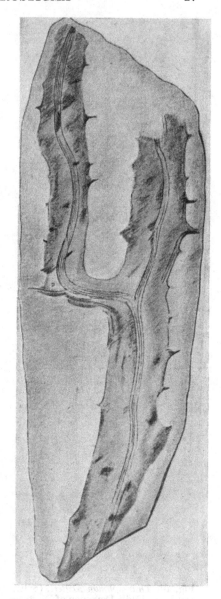

Fig. 8. *Arthrostigma gracile*, Dawson, from the Lower Old Red Sandstone of Scotland. (Nat. size.) After Kidston (1893).

Fig. 9.   *Arthrostigma gracile*, Dawson, from the Lower Devonian of Röragen, Norway. After Halle (1916).

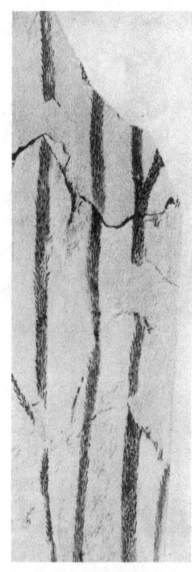

Fig. 10. *Thursophyton Milleri*, (Salt.) from the Middle Devonian of Western Norway. Sterile axes with scale-like emergences. After Nathorst (1915).

[1] Kidston (1893), p. 108.

thick, short (rudimentary or broken) distant leaves." The leaves here are pyramidal and thorn-like.

With regard to the morphology of these spine-like processes, Kidston[1] has concluded that these are not "the *bases* of leaves, as has been suggested, but are the leaves themselves, though developed in a very rudimentary form, as in *Psilophyton*."

Halle[2] says "the appendages of these stems can hardly be anything but leaves." To this we may reply that, on the present evidence, they are quite as likely to be *emergences*, especially in view of the similar structures occurring in certain species of *Psilophyton*, to which we have just drawn attention. On our view the emergences of *Arthrostigma* and *Psilophyton* differ chiefly in size and the variation in this respect in the former genus (which Halle has pointed out), as we have just said, is exactly what we should expect to find in view of the corresponding variation met with in *Psilophyton* (see pp. 21 and 22).

Halle also confirms Dawson's conclusion that the stem was vascular, the vas-

[2] Halle (1916), p. 13.

cular tissue forming a solid column without any pith. This much can be made out by macerating impressions.

Petrifactions of this genus are as yet unknown, nor do we know anything of its fructifications.

We think it probable however that *Arthrostigma* and *Psilophyton* are nearly related genera.

THURSOPHYTON.

(Figs. 10, 11.)

Thursophyton, Nathorst[1], 1915. Axes possibly herbaceous, dichotomously branched, of uniform thickness. Axes covered with crowded, imbricated, small (about 7 mm. long) scale-like emergences, spirally arranged and lanceolate in form, swollen at base, curving upwards. No leaf scars occur on the stem. Fertile shoots similar to the sterile axes bearing large (?) sporangia in the axils of some of the scale-like emergences.

Distribution. Upper Old Red, Scotland; Middle Devonian, Bohemia, Western Norway.

This fossil was originally described under the name *Lycopodites Milleri*, Salter, but, as Nathorst has pointed out, it has little or nothing in common with the Carboniferous plants referred to that genus and it is therefore best transferred to a new genus, *Thursophyton*. The first fertile shoots were described by Penhallow[2] and later by Reid and Macnair[3] as a distinct species *T. Reidi*, Penh. (Fig. 11, 1, p. 30). Here what appear to be globular sporangia, 1 mm. in diameter, frequently occur in the axils of some of the scale-like emergences. Nathorst[4], who has not seen actual specimens of these fertile shoots, regards Penhallow's interpretation of these fossils as fertile as being extremely doubtful. He suggests that it is not yet proved that the organs in question are sporangia and that they might be foreign bodies. While admitting the justice of some doubt on these points and the need of further evidence, we are inclined to think that Penhallow's interpretation will eventually prove to be in the main correct. For in the case of another species of the same genus,

[1] Nathorst (1915), p. 17.   [2] Penhallow (1892).
[3] Reid and Macnair (1896).   [4] Nathorst (1915), p. 19 footnote.

Thursophyton[1] (the *Lycopodites hostimensis* of Potonié and Bernard) (Fig. 11, 2) from the Middle Devonian of Bohemia, what are clearly sporangia occur in much the same way as in Penhallow's plant. The sporangia here are also large, 1·5 to

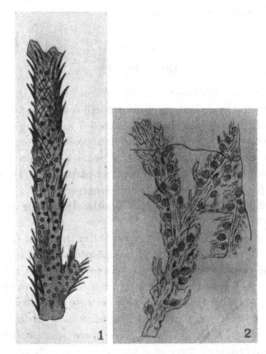

Fig. 11. (1) *Thursophyton Milleri*, (Salt.) = *T. Reidi*, Penh., from the Upper Old Red Sandstone of Scotland. A fertile shoot. After Reid and Macnair (1899). (2) *Thursophyton hostimense*, (P. & B.), from the Middle Devonian of Bohemia. Fertile shoots. After Potonié and Bernard (1904).

2 mm. in diameter, circular or perhaps a little elliptical. The stem in this case is dichotomous and the leaves are apparently broader and perhaps less crowded than in *T. Reidi*. We thus have very little doubt that the fertile shoots of *Thursophyton*

[1] Potonié and Bernard (1904), p. 45, Fig. 105 on p. 44.

are known and that their aspect is remarkably Lycopodian. The lateral organs, in view of the evidence of Psilophyton, we should be inclined to regard as scale-like emergences, though they may have begun to function as leaves. At any rate they do not on the present evidence appear to have been vascular.

Fig. 12. *Ptilophyton Thomsoni*, Daws. The lower part of a main axis from the Upper Old Red of Scotland. After Salter (1859).

PTILOPHYTON.

(Figs. 12–14.)

Ptilophyton, Dawson[1], 1878. Main axis (Fig. 12, p. 31) very stout, striated, (the *Caulopteris* ? *Peachii* of Salter[2]; cf. also the genus *Barrandeina*) giving off stout lateral decurrent shoots almost at right angles. Branches covered with scale-like emergences. At the apex the axis is freely and closely alternately branched, producing a tuft of shoots[3], the ends of which are circinately coiled (Fig. 13). The ramifications of this tuft bear, apparently on one side, a row of long thin (? filamentous) obscure organs, the nature of which is unknown. They have been described as "tufts of linear bodies[3]."

Distribution. Middle Devonian, ? Bohemia; Upper Old Red, Scotland.

This genus, markedly different from *Psilophyton* in habit, is still entirely obscure. We may note however that what is probably

Fig. 13. *Ptilophyton Thomsoni*, Dawson. The terminal portions of axes possessing emergences. Type specimen from the Upper Old Red of Scotland. (Reduced about ⅔.) After Carruthers (1873).

[1] Dawson (1878).
[2] Salter (1859), p. 407, Fig. 14 on p. 408; Kidston (1902) definitely states that *Caulopteris Peachii*, Salter, is the stem of *Ptilophyton Thomsoni*, Dawson.
[3] Carruthers (1873), Pl. 137.

Fig. 14. *Ptilophyton* (?) *hostimense*, (P. & B.). Type specimen from the Middle Devonian of Bohemia. After Stur (1881).

Fig. 15. *Pseudosporochnus Krejčii*, (Stur), from the Middle Devonian of Bohemia. Lower extremity of axis showing basal bulb. (Much reduced.) After Potonié and Bernard (1904).

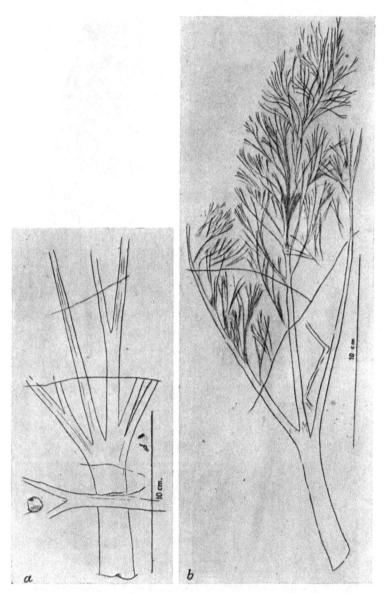

Fig. 16. *Pseudosporochnus Krejčii*, (Stur). (*a*) Branching of thallus (reduced). (*b*) Terminal portion of thallus (much reduced). After Potonié and Bernard (1904).

another representative of it (Fig. 14) occurs in the Middle Devonian of Bohemia and was referred by Potonié and Bernard[1] to the genus *Spiropteris*. These examples are again all equally obscure.

PSEUDOSPOROCHNUS.

(Figs. 15, 16.)

Pseudosporochnus, Potonié & Bernard[2], 1904. Axis stout and undivided below, bulbous? at base (Fig. 15), freely branched above (Fig. 16) in a pedate manner, secondary branches further dichotomised above, the slender branches of the third order being repeatedly and frequently dichotomised so that the higher parts of the secondary axes are clothed with fairly dense tufts of delicate, dichotomous, very narrow branchlets. The stems are known to be vascular.

Distribution. Middle Devonian, Bohemia.

This very remarkable plant is apparently only known from Bohemia. No fructification is described.

Fig. 17. *Bröggeria norvegica*, Nath., from the Middle Devonian of Western Norway. (Somewhat reduced.) After Nathorst (1915).

[1] Potonié and Bernard (1904), p. 11, Text-figs. 1–5 on p. 12.
[2] Potonié and Bernard (1904); Stur (1881).

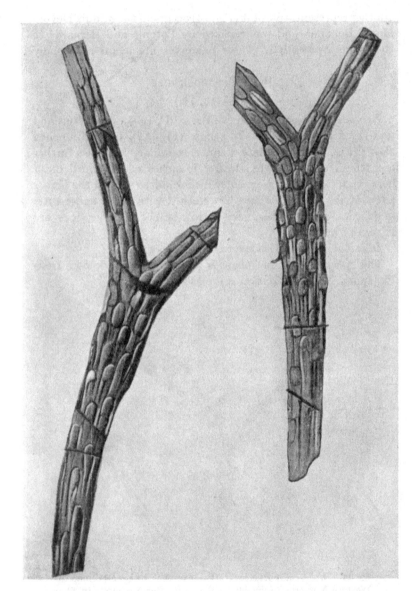

Fig. 18. *Barrandeina Dusliana*, (Kr.), from the Middle Devonian of Bohemia. (About ⅔ nat. size.) After Potonié and Bernard (1904).

BRÖGGERIA.

(Fig. 17.)

Bröggeria, Nathorst[1], 1915. Stout branched axes, of which some terminate in large cylindrical sporangial spikes or catkins, up to 50 mm. long and 15 mm. broad.

Distribution. Middle Devonian, Western Norway; ? Upper Devonian, United States.

The stems recall species of *Psilophyton*, but the fructifications are entirely dissimilar to anything known in that genus.

BARRANDEINA.

(Fig. 18.)

Barrandeina, Stur[2], 1881. Ribbed dichotomous axes; the ribs, which are formed by decurrent bases of lateral axes spirally arranged, are longitudinal, irregular, broad, low, flat, nearly contiguous, somewhat obscure.

Distribution. Middle Devonian, Bohemia, ? Western Norway and United States; Upper Devonian, United States.

No fructification is known in connection with these remarkable axes which appear to be largely made up of decurrent leaf bases. The lower portion of the axis of *Ptilophyton* (see Fig. 12, p. 31) appears to be somewhat similar and no doubt several of the specimens from the Devonian of America attributed by Dawson and Newberry[3] to the genus *Caulopteris* belong here.

BARINOPHYTON.

(Fig. 19.)

Barinophyton, D. White[4], 1905. Axes thick, smooth or irregularly ribbed, bearing alternate stout compact boat-shaped, fertile branches, usually lanceolate, consisting of a very thick fleshy keel, bearing on either side on its ventral surface, a row of alternating small thick•oblong or oblong-lanceolate scales or bracts. Bracts fleshy at the base, more or less distinctly carinate, provided with a small ventral pit or pocket, probably the seat of a sporangium.

---

[1] Nathorst (1915). [2] Stur (1881); Potonié and Bernard (1904).
[3] Newberry (1889). [4] White (1905).

Distribution. Upper Devonian of Belgium, United States, Canada and Australia.

Fig. 19. *Barinophyton Richardsoni*, (Daws.), from the Upper Devonian of the United States. After White (1905).

This fertile shoot is at present wholly obscure, but it is very unlike any other organ known from more recent rocks. It is widely distributed in Devonian rocks.

PARKA.
(Fig. 20.)

Parka, Fleming[1], 1831. Body small, of variable size, rarely exceeding one or two inches across, more or less circular, lenticular, of very small thickness, containing many disc-like oval or circular masses, which in their turn contain spores.

Distribution. Silurian and Lower Old Red, Scotland and England. Not known outside Britain.

Don and Hickling[2], in an important and quite recent paper on this mysterious fossil, have shown conclusively that the disc-like masses of *Parka* undoubtedly contain spores and that the fossil is thus clearly of vegetable origin. In this matter they confirm the conclusions of Dawson and of Penhallow[3] and more recently of Reid, Graham and Macnair[4]. Don and Hickling show that *Parka* consists of a flat, dorsiventral, multi-cellular and multi-layered thallus, developing by marginal growth (Fig. 20, 1). The shape is roughly circular or oval; but lobate, reniform and even irregular forms occur. In size the thalli measure 5 mm. to 7 cm. across. The margin, which is distinctly frilled, is usually less than 1 mm. broad. The rest of the thallus is composed of small oval

[1] Fleming (1831).   [2] Don and Hickling (1917).
[3] Dawson and Penhallow (1891); Penhallow (1892).
[4] Reid, etc. (1897) and (1899).

thickened discs of nearly constant size, usually 2 mm. in diameter, and of variable number. The discs are isolated spore masses, containing numerous cuticularised spores (Fig. 20, 3) of which there is no evidence that they were formed in tetrads, or that they were heterosporous. These discs never overlap, though they coalesce occasionally. A grooved lamina (Fig. 20, 2) occurs on one side of the thallus and is probably ventral.

In general habit Don and Hickling compare *Parka* with the

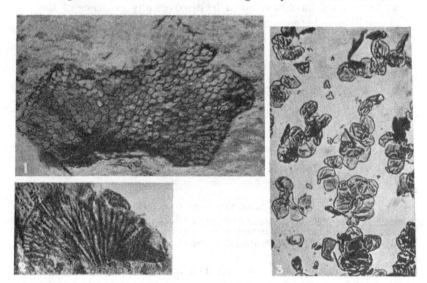

Fig. 20. *Parka decipiens*, Flem., from the Lower Old Red of Scotland: (1) a large thallus (natural size); (2) the folded lamina (×2); (3) spores (×150). After Don and Hickling (1917).

Coralline Alga *Melobesia* (*Lithophyllum*) *lichenoides*, Ag. of the Rhodophyceae (Fig. 21) and they conclude that the thallus grew probably attached to or on the surface of mud or sand.

With regard to the all-important question of affinity, Don and Hickling dissent entirely from the Dawsonian view that *Parka* was a sporocarp of "a somewhat generalised plant, shadowing forth the recent rhizocarps[1]." They regard it as a very low spore-bearing plant, belonging to a group which

[1] Reid, Graham and Macnair (1897).

possibly no longer exists, in fact as a "Thallophyte with Algal affinities[1]."

As to whether *Parka*, as here described, represents the whole plant, there must always remain a slight doubt, until petrified specimens are known. This uncertainty however appears to us to be very slight, and Don and Hickling find no evidence that *Parka* represented "aquatic plants with creeping stems, linear leaves" as had been asserted by Dawson and Penhallow[2] as late as 1891. They have also failed to discover any evidence of the prothalli and heterospores which the earlier workers believed they could recognise.

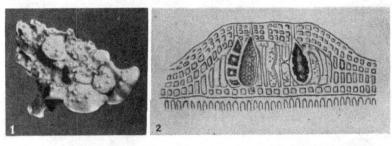

Fig. 21. Living Coralline Algae for comparison with *Parka*. (1) *Litho-phyllum lichenoides*, (E. & S.), external morphology (nat. size). (2) *Melobesia membranacea* (Esper) Lamour, vertical section through a conceptacle containing tetraspores (× 350). Both after Rabenhorst's *Kryptogamen-Flora* (1885).

There can be no doubt that Don and Hickling's work on *Parka* has removed a cloud of doubt and suspicion in regard to this fossil, much of which appears to have been ill-founded. Several points will remain uncertain until petrified material is discovered. All that has so far been made out of the structure of the thallus has been accomplished by means of macerating carbonised and cuticularised impressions. At the same time the important conclusion that we are dealing here with a lowly Thallophyte, not far removed from the Algae, and even comparable in habit to certain living Coralline Algae, is one which is not likely to be displaced, but rather to be confirmed, by a knowledge of structure material.

[1] Don and Hickling (1917), p. 661.
[2] Dawson and Penhallow (1891), p. 16.

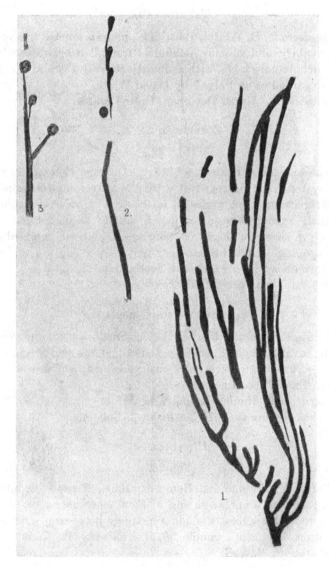

Fig. 22. *Zosterophyllum myretonianum*, Penh., from the Lower Old Red of Scotland. (Slightly reduced.) After Reid and Macnair (1899).

TAENIOCRADA.

Taeniocrada, D. White[1], 1905. This name is applied to a very fragmentary and obscure palmate "frond," deeply dissected into dichotomous lobes, with an indistinct central strand or axis. It is regarded as an "alga" by David White.

Distribution. Upper Devonian, United States.

ZOSTEROPHYLLUM.

(Fig. 22.)

Zosterophyllum, Penhallow[2], 1892. Obscure elongate ? axes aggregated in the form of a tuft, arising from a common horizontal ? axis; erect ? axes longitudinally finely striated; striations equal, parallel; axes ? ribbon-like, linear, simple or dichotomously branched. Some of the ? axes bore small, ? lateral, rounded or oval sporangium-like bodies.

Distribution. Lower Old Red, Scotland.

This fossil is at present wholly obscure.

PROTOLEPIDODENDRON.

Protolepidodendron, Krejči[3], 1879. Small leafy twigs with small, oval-lanceolate leaves; leafless stems with spirally arranged leaf bases; leaf bases small, fusiform; leaf scar absent or indistinct (? decorticated).

Distribution. Middle Devonian, Bohemia.

A very obscure type, *P. karlsteini*, in Bohemia.

HOSTIMELLA.

(Fig. 23.)

Hostimella, Potonié and Bernard[4], 1904. Repeatedly bifurcating slender branch systems without emergences, or other distinctive characters, though sometimes possessing circinate vernation. Typical example, *H. hostimensis*, P. & B. var. *rhodeaeformis*, P. & B.

[1] White (1905).
[2] Penhallow (1892); Reid and Macnair (1899).
[3] Krejči (1879); Potonié and Bernard (1904).
[4] Potonié and Bernard (1904).

Distribution. Lower Devonian, Norway (Röragen); Middle Devonian, Bohemia.

The examples of this genus are still entirely obscure objects. They appear to stand nearest to *Psilophyton.*

Fig. 23. *Hostimella hostimensis,* P. & B. var. *rhodeae-formis,* P. & B., from the Middle Devonian of Bohemia. (About half nat. size.) After Potonié and Bernard (1904).

APHYLLOPTERIS.

Aphyllopteris, Nathorst, 1915 = ? *Pteridorachis,* Nathorst, 1902. This genus, as emended by Halle[1], consists of stout rachis-like branch systems, without emergences, leaves or fructifications, and not dichotomously branched. As used originally by Nathorst, it included more slender dichotomously branched types such as *Hostimella.*

Distribution. Lower Devonian, Norway (Röragen); Middle Devonian, Western Norway; Upper Devonian, Bear Island and other countries.

[1] Halle (1916), p. 24.

This type is again wholly obscure and many examples of it no doubt represent small fragments of some of the preceding genera above discussed.

SPOROGONITES.

Sporogonites, Halle, 1916. Isolated stalked sporangia; sporangia obovoid or clavate, 6–9 mm. long and 2–4 mm. broad; apex rounded, base attenuated.

Distribution. Lower Devonian, Röragen, Norway.

This type has been recently described by Halle[1] from impressions from Röragen as examples of a Bryophytic sporogonium. He claims to have made out by maceration methods that the lower part of the capsule was "sterile throughout, the upper part consisting of three different zones: a wall of several layers of cells, a thick sporogenous tract and a sterile central columella[2]." The spores are tetrahedral, globular, 0·020–0·025 mm. in diameter, with cutinised walls.

An attentive examination of the description of these specimens given by Halle, has left us entirely unconvinced that any valid grounds exist for regarding these sporangia as sporogonia. In the absence of well-petrified material, it appears to us that the present distribution of the spores may well be secondary and not original. The fact that the wall of the sporangium, as we prefer to call it, is several layers in thickness has no bearing on the matter. The walls of the sporangia of *Psilophyton* (= *Rhynia*), as Kidston and Lang have shown, are also multi-layered, and to our eyes there is nothing about that genus which suggests affinities with the Bryophyta. The presence of a columella we regard as entirely unproven, and we doubt very much if the presence of such an organ, even if it undoubtedly existed, could be established from material preserved in the manner of the Norwegian specimens. That *Sporogonites* may be something more complicated than a sporangium with a simple uni-layered wall is quite possible, but even admitting this, it appears to us that, on the present evidence, its relationships are to be sought for

[1] Halle (1916).
[2] *Ibid.* p. 27.

among the Thallophyta rather than the Bryophyta. If the sporangium were septate, the longitudinal septa might be easily mistaken for a columella in material so imperfectly petrified[1].

[1] [These criticisms seemed justified at the time they were written, but the discovery by Kidston and Lang of a Middle Devonian plant, *Hornea Lignieri*, in which the sporangium undoubtedly possesses a columella, puts the whole question in a different light. The specimens of *Hornea* are petrified and the whole organisation of the plant is shown; the columellate sporangia are borne terminally on the dichotomous branches of a stem of the *Rhynia* type. While *Hornea* and *Sporogonites* are evidently quite distinct, there is now every reason to believe that Halle's interpretation of the structure of his fossil was essentially correct; the importance of his discovery is manifest, whatever view may be taken of the affinities of the plants in question. (See Kidston and Lang, On Old Red Sandstone Plants showing Structure, from the Rhynie Chert Bed, Aberdeenshire. Part II. Additional Notes on *Rhynia Gwynne-Vaughani*, Kidston and Lang; with Descriptions of *Rhynia major*, n.sp., and *Hornea Lignieri*, n.g., n.sp. *Trans. Roy. Soc. Edinb.* Vol. 52, Part III. 1920, p. 603.) D. H. SCOTT.]

CHAPTER IV

A DISCUSSION OF THE NATURE AND
AFFINITIES OF THE PSILOPHYTON FLORA

THE earlier conclusions as to the affinities of *Psilophyton* and
other members of that flora, advocated enthusiastically and
primarily by Dawson, and followed by some other workers,
need not detain us here. Dawson's[1] frequently repeated assertion
that *Psilophyton* and *Parka* in particular were related to the
Hydropterideae, or so-called Rhizocarps, was regarded by many
with grave suspicion, even at the time when no rival theory of
affinity was in the field. It is only necessary to add that all the
more recent work, especially the most recent studies of all
relating to these genera, has not produced, at any rate in our
opinion, one particle of evidence in favour of Dawson's conclusions as to affinity. In fact it may now be said that, whatever
views one may hold on this question, it is at any rate certain
that these plants were not related to the Water Ferns.

In any discussion of the affinities of these plants, the evidence
of *Psilophyton* must stand first. We know now the entire plant,
both in the form of impressions and petrifactions, and we are
thus in a singularly fortunate position where questions of
affinity are involved. We may first, however, state the views
of those who have quite recently contributed so greatly to our
knowledge in regard to this genus.

Halle[2] has no hesitation in regarding *Arthrostigma gracile* as
a microphyllous Pteridophyte, and he extends this conclusion
to *Psilophyton princeps*. He appears to base his conclusions
largely on the presence of a true vascular strand in these
plants.

Kidston and Lang have no doubt that "*Rhynia* and *Psilophyton*
belong to the Vascular Cryptogams or Pteridophyta[3]" and they

---

[1] Dawson (1888) and in many other places.
[2] Halle (1916).　　　　[3] Kidston and Lang (1917), p. 779.

propose to place them in a new class "the Psilophytales[1],"
"characterised by the sporangia being borne at the ends of
certain branches of the stem without any relation to leaves or
leaf-like organs[2]." Among existing Pteridophyta the authors
find in the living genera of the Psilotales the closest parallel to
Psilophyton.

These conclusions, however, should not be accepted without
some reservation. For our part we find ourselves unable to
adopt them, for it appears to us that *Psilophyton* has been
misinterpreted and that this and all the other genera belonging
to what has here been called the Psilophyton flora were much
more probably Thallophyta than Pteridophyta. This however is
likely to be a matter of prolonged controversy, involving a discussion of what we mean exactly by the former term.

In using this term here we recognise that the real problem is—
was *Psilophyton* simply a Thallophyte or was it a very reduced
Pteridophyte? In supporting the former view, as opposed to
recent workers on these fossils, we do not urge that, because this
or other genera were Thallophytes, they were necessarily Algae
in the sense in which that group is usually defined from a knowledge of its living members. On the contrary we think that
Psilophyton and some though perhaps not all the other genera,
belonged to a now obsolete race of Thallophyta, higher in the
scale of complexity than any living Algae. These plants we
propose to term the Procormophytes, and they will be further
discussed in a later chapter (p. 70).

Several members of the Psilophyton flora appear to have been
Algae pure and simple. This is the case with *Taeniocrada*
according to White[3] and *Parka* according to Don and Hickling[4].
On the other hand in *Psilophyton*, *Arthrostigma* and *Pseudo-
sporochnus*, we meet with other genera which appear to occupy
a somewhat higher position in the scale of morphological com-

[1] This term is open to considerable objections on the grounds that it is
too similar in form to another already in general use, i.e. the *Psilotales*.
If it is maintained, confusion is certain to arise from the similarity between
these names.
[2] Kidston and Lang (1917), p. 779.
[3] White (1905).
[4] Don and Hickling (1917).

plexity, particularly in the possession of a vascular strand in the main axis.

This anatomical fact, of the greatest interest and importance, seems to have mesmerized both Halle and also Kidston and Lang[1] to such an extent that the thought that *Psilophyton* might be a Thallophyte, does not appear to have ever occurred to them. The position is much the same as when Brongniart argued that *Sigillaria*, because it possessed secondary wood, must be a Gymnosperm! The possession of a vascular strand in the main axis does not appear to us to be a necessary sign of Pteridophytic affinity. Thallophyta are living to-day which possess a well-marked phloem. Thallophyta may have existed in the past which possessed a xylem strand. If they were terrestrial and not hydrophytes, it is highly probable that this was the case. *Psilophyton* was undoubtedly terrestrial. It is an immensely old type taking us back to days when terrestrial Algae of a high grade may have existed, though now long since extinct. At any rate it would be a very rash conclusion to deny that such plants have ever existed.

The Thallophyta are a race of plants which can by no means be kept within such narrow bounds. A race which among at least some of its members had evolved alternation of generations, a cormophytic habit, and true phloem, would if necessity arose be quite capable of evolving a lignified conducting tissue. If some of its members were at one time land plants, such a necessity would be obvious.

We have next to enquire whether *Psilophyton* presents any evidence of being an extreme case of Pteridophytic reduction rather than a Thallophyte. Here it appears to us that both the morphological and anatomical evidence is emphatically against the former view. If *Psilophyton* is such a reduced type, how is it that the xylem of the axis is reduced to a single protoxylem strand, a state of affairs unknown in any other, however highly reduced plant, whether living or fossil? How is it that a lack of

---

[1] [It should be noted that the present memoir was completed before the appearance of the following papers by Kidston and Lang: On Old Red Sandstone Plants showing Structure, from the Rhynie Chert Bed, Aberdeenshire, Parts II and III. *Trans. Roy. Soc. Edinb.* Vol. 52, 1920, pp. 603, 643. A. A.]

vascular connexion exists between the main axis and its branches? Surely here the evidence is emphatically on the side of primitiveness?

The morphological data appear to us to be equally emphatic. In the first place the fructification is wholly Thallophytic. The sporangia are quite unlike those of any fern borne on a highly reduced frond, and they find their nearest homologues among the Red Algae and those forms which possess a simple type of carpogonium. We have further no evidence at all of more than one type of reproductive organ, although the complete plant is undoubtedly known.

Further there is the evidence of the emergences, for such we believe to be the real nature of the lateral organs on the erect axes. These are anatomically non-vascular and histologically emergences and not branches, as their anatomy clearly shows. As we have pointed out here, these structures, in *Psilophyton*, are of varied size, micro- or macroscopic, and in the latter case scale-like. So far as we can see there are no grounds, either anatomical or morphological, for regarding these structures as leaves, however reduced. Yet this is the interpretation which must be put upon them if these genera are to be regarded as very reduced Pteridophytes. Finally the habit of *Psilophyton*, a rhizome giving off rhizoids, and erect naked axes, some terminating in sporangia, is much more typically Thallophytic than Pteridophytic.

We fail to find any ground of comparison except in habit (which taken alone is a perfectly valueless character) to the living plant *Psilotum*. The fructification and the vascular structure of the two are quite distinct. Further if, as we believe, *Psilotum* is rightly placed in close relation to *Tmesipteris*, it is obvious that any affinity between these plants and *Psilophyton* must be very remote.

We thus regard *Psilophyton* as first and foremost a Thallophyte, which, while still Thallophytic in habit, may occupy anatomically a place half-way between the Thallophyta and Pteridophyta. We propose to term such plants *Procormophyta*.

With regard to the other genera of the Psilophyton flora, the evidence is less emphatic, since they are less completely known.

But in many cases there is a strong family resemblance. Both *Arthrostigma* and *Thursophyton* possess abundant emergences, and also in habit are obviously not far removed from *Psilophyton*. The former is also known to be vascular. The habit of *Pseudo-sporochnus* (Figs. 15, 16, pp. 33, 34) is very remarkable and again we are dealing with a vascular type. The bulb at the base of the main axis (Fig. 15, p. 33) is an exceedingly algal feature, but in the upper part of the plant the finer branches assume a more or less Pteridophytic form (Fig. 16 *b*, p. 34), whereas the lower axes (Fig. 16 *a*) resemble Thallophytes. Stur[1] originally regarded this genus as an Alga pure and simple and even placed it in a living genus (*Sporochnus*, Ag.) of that group, but Potonié and Bernard[2] rightly point out that this genus is not a mere Alga, as its vascular structure clearly shows.

With regard to the other genera of the Psilophyton flora, *Ptilophyton*[3], *Hostimella*, *Bröggeria*, etc., we make no remark in this connection, save to point out that each is thalloid, and appears to have some features in common with *Psilophyton*, and that none of them are obvious Pteridophytes. They are however still far too imperfectly known to afford any secure evidence of the affinities of the flora to which they belong. *Zosterophyllum* is perhaps the most obscure of all, though as regards habit, it may be compared with species of the living Alga, *Nemalion*.

We shall discuss the bearing of the conclusions expressed here in regard to *Psilophyton* in a later chapter (p. 70).

In connection with the Psilophyton flora there remain one or two matters which require consideration. It might be urged that *Psilophyton* is a highly reduced type. The stomata are few and it might be thought that it is a reduced xerophyte. Against this view is to be set the clear anatomical evidence that the lateral organs are emergences and not reduced leaves. It is very unlikely that the latter would be microscopic, as we have seen they frequently are in *Psilophyton*. The absence of any leaf traces is again not what we should expect in the case of a leaf-less xerophyte. But probably the most convincing argument of

---

[1] Stur (1881), p. 342.   [2] Potonié and Bernard (1904).
[3] The terminal portion of this thallus is distinctly algal in appearance. Cf. *Procamium* spp.

all is that, if we regard *Psilophyton* as a reduced Cormophyte, all the other Procormophytes such as *Arthrostigma, Thursophyton*, etc. must be likewise regarded as reduced Cormophytes. This is clearly not the case and there is thus every reason to regard the Psilophyton flora as primitive and not reduced, especially as it can be shown that more highly evolved types sprang from them (see Chapter VI of this memoir).

Further, we know of no geological reasons which would lead us to suppose that the conditions of existence of plant life were at all different in Lower and Upper Devonian times. In the later epoch, members of the Psilophyton and Archaeopteris floras, as we have seen (pp. 9, 10), existed side by side, and were the latter in existence in Lower Devonian times there would seem to have been nothing to have prevented them from flourishing equally well at that period. As a matter of fact they did not then exist, for the incoming of the Archaeopteris flora is plainly indicated in Middle Devonian times, and it was clearly not established until the latest epoch of that period.

CHAPTER V

RECENT ADVANCES IN OUR KNOWLEDGE OF THE MORPHOLOGY OF THE ARCHAEOPTERIS FLORA

BEFORE discussing the phylogenetic deductions which, as it appears to us, may be drawn from a consideration of the Psilophyton flora, it may be well to review our present knowledge of the later or Archaeopteris flora of the younger Devonian rocks (see table, p. 9).

This flora is clearly *Cormophytic* and Pteridophytic but it sprang from the Psilophyton flora which as we have seen was *Thallophytic*.

In this case again we shall only notice the less obscure types, and in the case of genera which are well known and described in every text-book, we have not added any diagnoses.

SPHENOPSIDA.

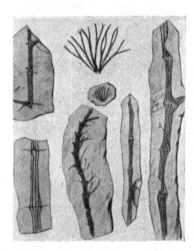

Fig. 24. *Sphenophyllum subtenerrimum*, Nath., from the Upper Devonian of Bear Island. Considerably reduced. After Nathorst (1902).

[1] Nathorst (1902).

Sphenophyllum, Brongn., 1828 (Fig. 24). Only a single species of this well-known genus is recorded from the Upper Devonian of Bear Island[1]. Like all the earlier representatives of the genus it has small, very narrow, highly divided leaves. The species with entire or nearly entire wedge-shaped leaves are not known earlier than the Upper Carboniferous. The Devonian type is very similar to, perhaps even identical with, a species occurring in the Lower Carboniferous.

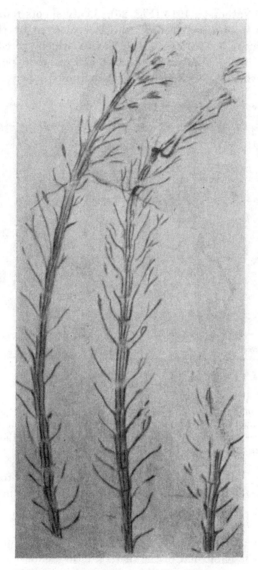

Fig. 25. *Hyenia sphenophylloides*, Nath., from the Middle Devonian of Western Norway. (Nat. size.) After Nathorst (1915).

Hyenia, Nathorst[1], 1915 (Fig. 25). Leafy shoots radiating in tufts from a stem or rhizome, shoots coalescent at the base. Shoots bearing leaves in whorls at nodes, which are either not prominent or very indistinct. Leaves at least four, perhaps to six in a whorl, successive whorls superposed. Leaves small,

Fig. 26. *Pseudobornia ursina*, Nath., from the Upper Devonian of Bear Island. (1) Leafy shoots (much reduced). (2) Fertile shoots (greatly reduced). Both after Nathorst (1902).

10–15 mm. long, rarely 20 mm. long and 1 mm. broad, uninerved, forked once or more rarely twice at apex. Fructification unknown.

Distribution. Middle Devonian, Western Norway.

This type appears to be near to *Sphenophyllum* but perhaps differs in the habit and in the fact that the nodal lines are not evident.

Pseudobornia, Nathorst[2], 1894 (Fig. 26). Stems large, seg-

[1] Nathorst (1915).   [2] *Ibid.* (1894).

Fig. 27. *Psygmophyllum Kolderupi*, Nath., from the Middle Devonian of Western Norway. Leaves and leafy shoots. (Nearly nat. size.) After Nathorst (1915).

mented, nodes curved, irregularly branched. There may be 1–2 branches on each node. Leaves shortly stalked, borne in whorls of four at the 'node, dichotomously divided, margins finely toothed, veins fan-like. Fructification a catkin-like body, up to 32 cm. long, with short internodes; sporophylls whorled,? forked dichotomously, ? all fertile in the lower parts; sporangia containing megaspores.

Distribution. Upper Devonian, Bear Island.

This very interesting type is a now well-acknowledged member of the Sphenopsida. It is a very rare plant.

Fig. 28. *Archaeopteris hibernica,* (Forbes), from the Upper Old Red of Kiltorkan, Ireland. (1) Complete frond (greatly reduced). (2) Fertile portion of a frond (much reduced). (3) Sterile pinnules (reduced). After Carruthers (1872).

PALAEOPHYLLALES[1].

Psygmophyllum, Schimper, 1870 (Fig. 27, p. 55). Leaves large, flabellate or cuneiform, arranged spirally on an axis, sheathing at the base, which is fairly broad and not contracted to a slender petiole. Apex broad, rounded or truncated, entire or lobed, or slightly divided. Nerves flabellate. Fructification entirely unknown.

Distribution. ? Lower Devonian, Spitzbergen; ? Middle Devonian, Western Norway; Upper Devonian, Ellesmereland, Spitzbergen, Canada and United States; Lower and Upper Carboniferous.

This is a very striking and widely distributed genus in Devonian rocks. A monograph of it was published by the

[1] Arber (1912), p. 405.

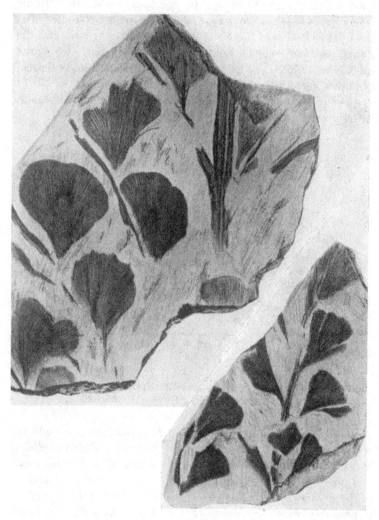

Fig. 29. *Archaeopteris Archetypus*, Schmalh., from the Upper Devonian of Ellesmereland. (Nat. size.) After Nathorst (1904).

writer[1] some years ago. The large wedge-shaped leaves are in most cases detached, except in specimens from the Middle Devonian of Western Norway, and much younger examples from the Coal Measures of the Newcastle coal field, England, and the Permo-carboniferous of South Africa, where they are , borne spirally on a shoot. The Devonian and the British (Carboniferous) examples agree remarkably in habit in this respect.

Although several species of the genus are known, we remain entirely unacquainted with the fructification.

Fig. 30. *Archaeopteris fimbriata*, Nath., from the Upper Devonian of Bear Island. (⅔ nat. size.) After Nathorst (1902).

It has been thought probable that *Psygmophyllum* belongs to a distinct group of plants, for which the name *Palaeophyllales* was suggested by the writer some years ago. Whether this view will prove to be correct or not, depends primarily on the nature of the unknown fructification. Until this has been discovered it is probably wiser to assume that it represents a distinct type than to include it with any other Palaeozoic genus, for among such plants the habit of *Psygmophyllum* is quite unique.

[1] Arber (1912). To complete the lists contained in this monograph the figures of *Platyphyllum Brownianum* of D. White (1905), of *P. obtusa* by Prosser (1894), Pl. II, and those of Nathorst from Western Norway, Nathorst (1915), should be added. It is now admitted that the Permian types from Russia at one time included in this genus are in reality quite distinct.

PTEROPSIDA.

Archaeopteris, Dawson, 1871[1] (Figs. 28–32). Fronds of large size, bipinnate with a stipular base, stipules in pairs, adnate, and a ramentum on the lower part of the petiole. Sterile pinnules typically obovate or ovate-cuneate, entire or toothed, with a flabellate

Fig. 31. *Archaeopteris fissilis*, Schmalh., from the Upper Devonian of Elles-mereland. (Nat. size except fig. marked 9, which is × ⅔.) After Nathorst (1904).

[1] See Carruthers (1872); Kidston (1888); Nathorst (1902) and (1904); White (1905); Kidston (1906), p. 434; Johnson (1911[2]).

nervation, or less typically the pinnules are lobed, fimbriated or divided longitudinally into numerous narrow spreading segments. Sterile pinnules occur on the main axis between the insertion of the pinnae. Fertile pinnules [1] very reduced, occurring on the same frond as the sterile leaflets, usually the whole or part of the lower pinnae of the frond being fertile. Fertile

Fig. 32. *Archaeopteris Hitchcocki*, (Daws.), from the Upper Devonian of United States. A fertile frond. The type specimen (nat. size) after White (1905).

pinnules bearing single or grouped (2–3) sporangia, sessile or shortly stalked, sporangia fairly large, fusiform or oval, exannulate.

Distribution. Upper Devonian to Middle Coal Measures. The affinities of this genus will be discussed at a later stage (p. 81).

[1] Sporangia may also occur on the margins of pinnules similar to sterile leaflets, Kidston (1906), p. 434, and Johnson (1911²).

Rhacopteris, Schimper, 1872 (Fig. 33). Fronds pinnate or dichotomously branched. Pinnules large, unsymmetrically wedge-shaped, rhomboidal, typically entire or more or less deeply lobed or divided longitudinally, with a radiating unsymmetrical nervation. Higher part of the frond sometimes fertile, sporangia tufted, small, exannulate, globular.

This genus is very rare in the Devonian, though it appears to occur on that horizon in Germany. It is more characteristic of the Lower Carboniferous.

Distribution. Upper Devonian to Middle Coal Measures.

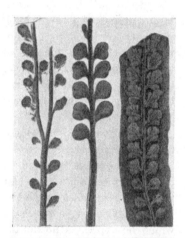 

Fig. 33. *Rhacopteris furcillata*, (Ludw.), from the Upper Devonian of Germany. (About ⅓ nat. size.) After Potonié (1901).

Fig. 34. *Sphenopteridium rigidum*, (Ludw.), from the Upper Devonian of England and Germany. (About ⅓ nat. size.) After Potonié (1901).

Sphenopteridium[1], Schimper, 1874 (Figs. 34, 35). An indefinable generic name applied to a particular type of Sphenopterid frond in which the pinnules are highly divided into very narrow

---

[1] This term is to be preferred to *Rhodea*, Presl, 1838, since that term is preoccupied for Angiosperms (*Rhodea*, Endlicher, 1837; *Rohdea*, Roth, 1821).

linear or filiform forked lobes as in the type species *S. rigidum*,
Ludw. (Fig. 34).

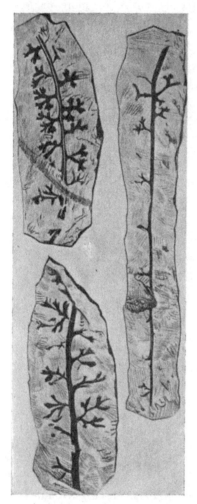

Fig. 35. *Sphenopteridium Keilhaui*, Nath., from the Upper Devonian of Bear Island. (Nat. size.) After Nathorst (1902).

Distribution. Middle Devonian to Upper Carboniferous.

Cephalopteris, Nathorst[1], 1910 (Fig. 36). Axes branched, branches opposite in distichous pairs, each pair connate on one side of the axis and decurrent. Lower portions of branches fertile. Sporangia in dense spherical heads arising from the decurrent base of each lateral branch. Sporangia long, pointed, dehiscing ? longitudinally. Sterile foliage of the ? Sphenopteridium type, upper portions of branches bearing small dichotomised leafy segments.

Distribution. Upper Devonian, Bear Island.

If Nathorst is correct in correlating certain sterile leaf segments (Fig. 36 (5)) with the fertile main axes—a point on which he expresses no doubt, though to us there seems to be no proof beyond mere association—then this type is simply a fertile *Sphenopteridium*, unless in its fructification it is distinct from other members of that genus, the fructifications of which are at present quite unknown. For the

[1] Nathorst (1902), first described under the name *Cephalotheca* in that year.

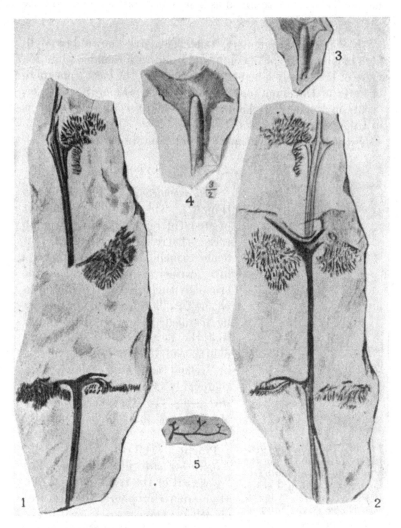

Fig. 36. *Cephalopteris mirabilis*, Nath., from the Upper Devonian of Bear
Island. Figures 1–4, fertile shoots (1–3 nat. size, 4 × $\frac{3}{2}$). Figure 5, sterile foliage
(nat. size). After Nathorst (1902).

moment it may be retained as a distinct type in a position close to *Sphenopteridium*.

Sphenopteris, Brongniart, 1822. This well-known but well-nigh indefinable type of compound frond with rounded pinnules, more or less deeply lobed and contracted at the base, appears to be rarer in Devonian rocks than *Sphenopteridium*. It, however, occurs in England, Ireland, Belgium and several other regions in Upper Devonian rocks.

Distribution. From Upper Devonian onwards.

LYCOPSIDA.

Bothrodendron, L. et H., 1885 (Fig. 37). This well-known genus is of frequent occurrence in Devonian rocks. There has been some tendency to include the Devonian species in a distinct genus *Cyclostigma*[1], as originally suggested by Haughton in 1859. This, however, can now hardly be justified[2]. There is little doubt that the best known of the Devonian species, *B. Kiltorkense*, occurring in Ireland and Bear Island, is a thoroughly typical representative of the genus, as the recent studies of Nathorst and Johnson clearly show.

Fig. 37. *Bothrodendron Kiltorkense*, (Haugh.), from the Upper Old Red of South Ireland. Stem with leaf scars. (Reduced ½ nat. size.) Specimen No. 20 Devonian Plant Coll., Sedgwick Museum, Cambridge. (W. Tams photo.)

Practically all the organs of *B. Kiltorkense* are now known. The lower part of the trunk consists of a Stigmarian rhizophore, the features of which agree closely with the Stigmarias of the Coal Measures.

According to Johnson[3], "the leaves are clearly arranged in whorls at first, but become distant and quincuncially arranged in older stems, owing to

[1] This term is in any case several times preoccupied by recent Angiosperms.
[2] Johnson (1913), especially p. 505.   [3] *Ibid.* (1913), pp. 523–4.

the unequal extension of the stem surface. ...The [decorticated][1] stem may show a marked fluting or ribbing which is connected with the parichnos and bundle-strands, but possibly also with internal sclerotic bands. The calamitoid appearance of such stems is increased by the presence of horizontal or transverse ridges or zones which are, unlike the longitudinal ridges, coincident with the surface leaf-scars and suggestive of nodal diaphragms....The [heterosporous][1] cone is terminal, and carried on its broad (hollow?) axis numerous whorls of sporophylls, of which the megasporophylls are the ones at present best known." The sporophylls appear to be of a leafy type well known in *Lepidostrobus*, bearing sporangia on the upper surface of the basal portion. The distribution of mega- and microsporophylls in the cone is at present unknown.

With regard to the calamitoid appearance of certain decorticated stems of this genus, we differ from Johnson who is inclined to see in this feature some signs of affinity to the Sphenopsida. These specimens appear to us to represent sub-epidermal surfaces which are not comparable with Calamite pith casts. Further in many, but not all, Calamites, the external or sub-external surface of the stem was not ribbed longitudinally[2].

Neither the leaves, which are uninerved, long, linear structures of the usual Lycopod type, nor the cones attributed to *B. Kiltorkense* have as yet been found attached to the stems. The former are believed to have been caducous.

Distribution. Upper Devonian to Upper Carboniferous.

Archaeosigillaria, Kidston, 1901[3] (Figs. 38, 39, p. 66). Plants with stems attaining a diameter exceeding 2·5 cm., dichotomously branched. Stem covered with spirally arranged persistent leaf bases; leaf bases contiguous, fusiform in younger branches, hexagonal in older stems, bearing a single print situated slightly above the centre of the leaf base. Leaves small, deltoid, markedly

---

[1] Inserted by the present author.

[2] At the same time there is no doubt that the occurrence of these ribbed stems has given rise to the assertions of Heer and others that such genera as *Calamites* or *Archaeocalamites* occur in Devonian rocks, of which, however, there is no real evidence.

[3] For typical figures see Kidston (1885) and White (1907).

falcate. Detailed structure of cone unknown. This genus is now fairly well known from a species occurring both in the Lower Carboniferous of England and in the Upper Devonian of America.

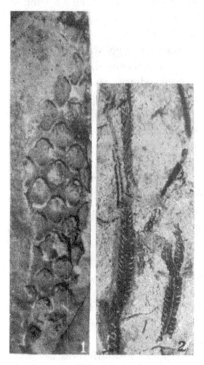

Fig. 38. *Archaeosigillaria Vanuxemi*, (Goepp.), from the Upper Devonian of the United States, and Lower Carboniferous of England. (1) Leaf scars enlarged ⅔. (2) Leafy twigs (nat. size). Specimens Nos. 1099 and 1106, Carboniferous Plant Collections, Sedgwick Museum, Cambridge. (W.Tams photo.)

Fig. 39. *Archaeosigillaria primaeva*, White, from the Middle Devonian of the United States. Stem with leaf scars (reduced). After White (1907).

In the latter country it also occurs in Middle Devonian beds[1]. It is also probable that, as White[2] has suggested, certain North American fossils ascribed to *Lepidodendron* really belong here.

[1] White (1907).   [2] *Ibid.* (1907), p. 339.

Distribution. Middle Devonian to Lower Carboniferous.

In the specimens from the Middle Devonian of the United States, some of the decorticated stems are markedly ribbed, just as we have seen is the case in Devonian Bothrodendrons.

Leptophloeum, Dawson, 1862[1] (Fig. 40). Stems subarborescent, dichotomously branched, with a spirally arranged armour of leaf bases; leaf bases of relatively large size as compared with the diameter of the stem, nearly contiguous, rhomboidal, arranged in periods of large rhomboidal bases alternating with periods of much smaller, more transversely elongated bases. Leaf scar very small, situated a little above the middle of the base, oval or ovate, with a single print situated a little above the middle of the scar. Ligular ? pit at apex of base.

We agree with White[2] in referring the so-called *Lepidodendron australe,* McCoy, and *L. nothum,* Unger, of Australia, to this genus.

Distribution. Devonian, only in Canada, United States, Spitzbergen and Australia.

Fig. 40. *Leptophloeum rhombicum,* Daws., from the Upper Devonian of the United States. Type specimen (nat. size). After White (1905).

ISOLATED FRUCTIFICATIONS OF UPPER DEVONIAN AGE.

The most common types of isolated fructifications occurring in Upper Devonian rocks are those which are clearly similar to, or even generically identical with, the fructifications of species of *Archaeopteris.* The few other types found, including *Xenotheca*[3] in England and *Dimeripteris*[4] of Russia and the United

---

[1] White (1905).
[2] *Ibid.* (1905), pp. 72–3.
[3] Arber and Goode (1915), p. 96, Pl. IV, figs. 1–7, 10, 11, Text-fig. 2.
[4] White (1905), p. 53.

States, the latter having something in common—though perhaps remotely, with the Crossothecas of the Coal Measures, are at present wholly obscure.

The absence of seeds associated with the undoubted Devonian floras is very remarkable. A revision of this flora has not produced a single undoubted specimen of a seed. The three supposed examples attributed to the genus *Carpolithes* by Dawson in 1863 and derived from the Devonian of America, have no claim whatever to be regarded as seeds[1]. One of these more recently figured by White is a small object of a doubtful nature which he thinks may be merely a scale, possibly comparable to those of *Barinophyton*[2].

The present author has seen from the Kiltorkan beds of South Ireland, one or two small bodies which bear some slight resemblance to seeds, but it is quite possible that they may be capable of an entirely different explanation. If seeds do occur at Kiltorkan, they are undoubtedly exceedingly rare, and since the number of species of all groups there represented is very small, probably not more than four, it is unlikely that *Archaeopteris*, which is there by far the commonest type represented, is a seed-bearing plant. This is a point to which we shall return later (p. 81). For the present it is best to assume that *undoubted* seed impressions are unknown from Devonian rocks.

GENERA OF VERY DOUBTFUL OCCURRENCE IN DEVONIAN ROCKS.

Lepidodendron, Sternberg, 1820. There is still no really convincing evidence of the occurrence of this well-known genus in Devonian rocks, from which, however, it has been reported from many parts of the world. That is to say, no examples have been figured which show a typical Lepidodendroid *leaf-scar* and its prints. If the genus does occur, the specimens so far known appear to be all more or less decorticated. This is obviously the case in some examples figured from Australia, Canada, the Arctic Regions, Russia, etc. It seems probable, however, that some of the best preserved of these fossils belong to distinct genera such as *Archaeosigillaria* and *Leptophloeum*. The type

---

[1] Cf. White (1905), p. 78.   [2] White (1905), p. 78, Pl. IV, fig. 11.

which occurs widely and frequently in Australia, the so-called *Lepidodendron australe*, is best referred to *Leptophloeum*, which is wholly Devonian. White[1] has also recently concluded that by far the greater number of the so-called American Lepidodendreae belong in reality to *Archaeosigillaria*, occurring both in the Upper Devonian and Lower Carboniferous though not in later rocks. *Lepidodendron* appears to be wholly Carboniferous and Permian.

Cordaites, Unger, 1850. The occurrence of Cordaites-like leaves in Devonian rocks is very rare. One such is known from England[2], though in a fragmentary condition, and others have been recorded from Australia, but it cannot be said that the evidence for the existence of *Cordaites* in Devonian times is at present at all trustworthy. The evidence from the Devonian of America is even less satisfactory.

Distribution. ? Upper Devonian, Lower Carboniferous to Permian.

Genera unknown in the Devonian Rocks.

Despite many assertions to the contrary in the older literature, the following genera do not appear to be known in Devonian rocks, or rather there are no trustworthy records of their occurrence in those beds.

These genera were mostly well developed in Lower Carboniferous times, as was also the genus *Rhacopteris* which is exceedingly rare in the Devonian. Thus the Lower Carboniferous flora is distinguished from that of the Devonian, by the presence of the following genera, in addition to others common to the two formations.

Equisetales	{ *Archaeocalamites* { *Calamites* (rare)
Pteridospermae or Filicales	} *Adiantites* } *Cardiopteris*
Lycopodiales	{ *Lepidodendron* { *Lepidophloios* { *Sigillaria* (rare)
Cordaitales	*Cordaites* (rare)

[1] White (1907).
[2] Arber and Goode (1915), Pl. V, fig. 5.

CHAPTER VI

THE PROCORMOPHYTA AND THE ORIGIN OF THE CORMOPHYTA (EXTERNAL MORPHOLOGY)

HISTORICAL.

THAT these recent discoveries affecting the Psilophyton flora have a very important theoretical bearing has been apparent to all workers on the subject. Halle[1] has recently entered into these matters at some length and a further discussion is promised by Kidston and Lang[2]. The origin of Cormophyta is a problem which has already attracted considerable attention, chiefly on the basis of a study of recent plants. Bower's *Origin of a Land Flora* (1908) contains an excellent epitome of these conclusions. In addition there have been studies in which fossil plants of Devonian age, and especially *Psilophyton*, have played some part. The earliest of these was apparently Potonié's[3] theory of the descent of all Pteridophyta from Algal ancestors[4], a view which we are inclined to support enthusiastically.

Potonié[5] supplemented this conception by his pericaulome theory, to which we shall refer again shortly. Both these explanations appear to us to be on the right lines, although we suggest a more limited application in regard to the pericaulome theory than that which Potonié himself advocated.

Almost simultaneously with Potonié's publications, a quite independent series of conclusions were put forward by the late Prof. Lignier. In 1903 Lignier[6] postulated a primitive terrestrial type, the Prohepatics, as having given rise to the Vascular Cryptogams on the one hand and to Bryophyta on the other.

[1] Halle (1916), p. 35. [2] [See footnote, p. 48. A. A.] [3] Potonié (1898).
[4] For a recent discussion of the same subject from the point of view of a study of recent Algae see Fritsch (1916). [In this connexion see also Church, A. H., *Thalassiophyta*, Oxford University Press, 1919—a memoir which had not appeared when the present book was written. A. A.]
[5] Potonié (1898), p. 19, (1902[1]) and (1902[2]).
[6] Lignier (1903) and (1908).

On this view the Equisetales and Sphenophyllales are descended from a common fern-like ancestor[1]. Lignier recognised *Psilophyton* as a primitive Vascular Cryptogam[2]. In his chief contribution to this subject[3], which appeared in 1909, he elaborates his views so as to cover the whole vegetable kingdom. From aquatic Algal ancestors, he derives terrestrial Prohepatics, giving rise to Bryophyta on the one hand and Prolycopods on the other. From the latter are derived the Primofilices and Lycopods. In the development of these two races[4], two distinct morphological tendencies are recognised by Lignier as at work:

1. Phylloideae: leaves originating from emergences of the thallus— Prolycopods, and Lycopods alone.

2. Phyllineae: leaves originating from modified thalloid branches— Primofilices, and all Vascular Cryptogams except Lycopods, and all Spermophyta.

The former were *ab initio* microphyllous, the latter megaphyllous. The Sphenopsida or Articulatae are regarded as derived from megaphyllous fern-like ancestors. The author also proceeds to consider certain Spermophytes such as the Coniferae which are microphyllous, but as we are not here concerned with any group higher than the Vascular Cryptogams, these matters need not detain us.

From this view we should be inclined to dissent in several particulars. We regard it as extremely unlikely that the Bryophyta, using that term in the widest sense, had any connection with the origin of Pteridophyta. We should agree rather with Fritsch[5], in deriving the Bryophyta independently from the Algae but at a much later period than that with which we are here concerned. Halle's *Sporogonites* (see p. 44), which we have discussed, has not shaken our conclusions in this respect.

On the other hand we are strongly in favour of accepting Lignier's conceptions of the Phylloideae and Phyllineae, though on somewhat different lines, but at the same time we dissent from any notion of either a primitive fern or prolycopod ancestry for the Sphenopsida and Pteropsida.

---

[1] Lignier (1903), p. 132.   [2] *Ibid.* (1903), p. 95.
[3] *Ibid.* (1909); see also Scott (1910).
[4] Cf. Janchen (1911) who also separates the Lycopods from other Pteridophyta on phyletic grounds.   [5] Fritsch (1916).

Our next historical reference, founded chiefly on the palaeobotanical evidence, is to the view of Dr Scott[1] published
some years ago in regard to the descent of Pteridophyta along
three main lines: the Sphenopsida, Pteropsida and Lycopsida.

Lastly we have Halle's[2] recently expressed conclusions founded
on the evidence of the *Psilophyton* flora. Apart from his views
on the antiquity of the sporogonium and Bryophyta to which
we have already referred (p. 44), Halle regards *Arthrostigma* as
a microphyllous Pteridophyte possessing leaf-bearing stems and
Psilophyton as possibly of the same nature. Our interpretation
is that they are Thallophytes possessing emergences. We, however, agree with Halle in regarding the vascular structure as
primitive in these genera.

Halle also discusses the question whether megaphyllous
Pteridophyta occurred in Lower Devonian times.

With this brief review of previous opinion, we proceed to
point out how recent work on the Devonian floras appears to
us to establish even more securely the conception of the early
existence of three distinct lines of descent, the Sphenopsida,
Pteropsida and Lycopsida. In the present chapter we shall
confine our attention to questions concerning the origin of the
external morphology of these groups.

THE CORMOPHYTIC HABIT.

We regard it as probable that the Psilophyton habit (Figs. 1
and 2, pp. 15 and 16) was primitive for all three lines of Cormophytic descent. That is to say, there was an epiterraneous or subter-
raneous limited, erect or horizontal axis, fixed by rhizoids or roots,
and giving off one or more vertical erect branches. This type of
habit persists throughout the earlier Sphenopsida (e.g. *Calamites*),
Pteropsida (*Pteridosperms*) and Lycopsida (*Lepidodendron*).

These three groups, however, differed essentially as regards
the morphology of the erect shoots and leaves, and it was the
evolution of these distinct lines of modification of types met
with among the Psilophyton flora—as is already foreshadowed
in the Archaeopteris flora—that originally called these three
groups into being. In some cases the main erect axis was trans-

---

[1] Scott (1909), p. 616, and (1910).   [2] Halle (1916), pp. 35–40.

formed into the aerial shoot directly, and was not built up partly of decurrent branches. This was the case in the Sphenopsida. On the other hand in the Pteropsida, and to a less general extent in the Lycopsida, the erect axes possessed a pericaulome, as Potonié[1] has shown, the external tissues being morphologically lateral organs fused to the original axis. The same thing of course occurs in many Algae, e.g. *Polysiphonia*.

The leaves also vary in origin. In the Sphenopsida they are metamorphosed small lateral branch systems, and, in the Pteropsida, large lateral branch systems. In Lycopsida they are metamorphosed emergences.

We should define these lines of evolution as follows:

(1) *Sphenopsida*, descended from Thallophytic Algae bearing whorled branches. Limited, whorled branches typically small, converted into leaves which were originally and always microphyllous. Stem not built up of foliar decurrences. Sporangia and sporangiophores, modifications of branches or segments of the same.

(2) *Pteropsida*, descended from Thallophytic Algae in which the branches were large, numerous, scattered and not whorled. Branches compound, eventually metamorphosed to megaphyllous leaves. Stem largely built up of foliar decurrences (pericaulome). Sporangia and sporangiophores, modifications of segments of branches.

(3) *Lycopsida*, descended from Thallophytic Algae in which the aerial axes were rarely branched and then usually in a dichotomous manner. The branches bore microscopic or macroscopic emergences which were metamorphosed to microphyllous leaves. Stem partly built up of foliar decurrences in some cases. Sporangia and sporangiophores, modifications of emergences or segments of emergences.

The above outline agrees exactly with Potonié as regards Algal ancestry, but to a limited extent only as regards pericaulome characters. It agrees with Lignier as regards the

---

[1] Potonié (1898), p. 19, (1902[1]), (1902[2]). The pericaulome theory has more recently been disputed by Kubart (1913) on anatomical grounds. In the present paper, however, it is used solely as a morphological conception without reference to the physiological functions which the fused organs may be supposed to have performed originally.

Lycopsida—his Phylloideae—but in place of his Phyllineae, two distinct lines of descent are recognised, and the Sphenopsida are regarded as primitively microphyllous and not megaphyllous. In this and other features these conclusions differ from any previously expressed, and since our interpretation of *Psilophyton* and *Arthrostigma* differs from that to which Halle has inclined (though in some cases he admits that we may here be dealing with emergences and not leaves) so our conclusions as regards phylogeny are quite distinct from his.

We now propose to trace the early stages of the evolution of each group in some detail.

The Evolution of the Sphenopsida.

It seems probable that the Sphenopsida took their origin from Algae possessing a whorled habit like that of the living fresh water Red Alga—*Batrachospermum*. In this ancestor we should expect to find some form of primitive vascular system, at least as far advanced as in *Psilophyton*, and a distinct alternation of generations.

As regards the vegetative habit, we should expect to find that the whorled branches gradually took on the leaf function, without any very radical change in external morphology. They remained either simple or dichotomously forked microphyllous elements, the branching of the primitive lamina being the most general method of increasing the lamina area, in view of the new functions which these organs had assumed. As regards the cones, the primitive whorled habit was again retained. These were originally made up of whorled branches, the branches being sometimes simple and sometimes forked several times. Some of the branches of the whorl or their segments took the path of remaining protective organs (i.e. the bracts) and in some cases became coalescent to a considerable degree in order to fulfil this function more effectually. Other segments were metamorphosed partly into sporangiophores, partly into sporangia.

Such represents our conception of the main course of evolution in this group. We now have to enquire what fossil evidence we have for the support of these contentions.

With regard to the primitive whorled algal type there is little evidence—perhaps no certain evidence—at present. Attention may however be drawn to some very obscure fossils which may conceivably have some bearing on this point. At the same time it must be freely admitted that these bodies are extremely obscure and that no emphasis can be laid on them from this point of view. There do, however, occur in the Ordovician, Silurian and Devonian rocks, carbonaceous impressions which in some respects recall the Annularian leaves of Calamites. They

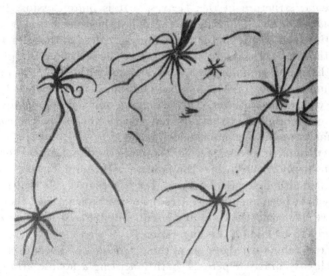

Fig. 41. *Protannularia laxa*, (Daws.), from the Upper Devonian of Canada ($\frac{2}{3}$ nat. size). After Dawson (1871).

have in fact more than once been referred to *Annularia* itself. Perhaps the best specimens of this nature are those described by Dawson[1] from the Devonian of Canada as *Annularia laxa*, Daws. (Fig. 41).

Dawson[2] states that "the ends of the leaves are curled in a circinate manner" and he figures them as each possessing a vein. It may be also recalled here that many years ago Nicholson[3]

[1] Dawson (1871), p. 31, Pl. VI, fig. 64; cf. also figs. 65–69.
[2] *Ibid.* (1871), p. 31.
[3] Nicholson (1869), p. 495, Pl. 18, fig. B; Nicholson and Lydekker (1889), p. 1514.

also figured from rocks as ancient as the Lower Ordovician (Arenig) of the Lake District, similar specimens under the name *Protannularia (Buthotrephis) radiata*, (Nich.). Examples of these are in the Sedgwick Museum, Cambridge[1]. Nicholson[2] says "it is...difficult...to imagine what this can be if not a plant.... It seems, however, pretty certain that if its vegetable nature be conceded, it can hardly be referred to the Algae." With this conclusion we entirely agree, though we find no trace of any vascular structure in the "leaves." It is quite possible that *Protannularia radiata* may be the oldest, in a geological sense, British land plant, and Procormophyte. Most authorities will we think agree that these specimens have no real claim to a place in the genus *Annularia*. We may transfer them to the non-committal place of *Protannularia*. It may be objected that these specimens are mere *lusus naturae*, mineral aggregates of inorganic origin, but this does not appear to us to be the case, so far as we have examined such fossils.

We think it probable that the earlier examples may have been simply Algae[3], while the later were Procormophytes—still thalloid although vascular—and that it is conceivable that they may have been among the ancestors of the Sphenopsida, though as we have said they are at present still too obscure for any weight to be laid on this suggestion.

From some such ancestors as these, two lines of descent sprang at slightly different periods, both inheriting a nodose arrangement of microscopic branches. The branches became vascular gradually, in exactly the same way as those of *Psilophyton* became vascular.

In the older line of descent, the Sphenophylls, the branches were characteristically divided longitudinally into forked segments. This is well seen in the leaves of all members of this group known in the Archaeopteris flora, i.e. *Sphenophyllum* (Fig. 24, p. 52), *Hyenia* (Fig. 25, p. 53), and *Pseudobornia* (Fig. 26, p. 54). These organs remain morphologically thalloid. The wedge-shaped

---

[1] The type No. 51 (also No. 1), Ordovician Plant Coll., Sedgwick Mus., Cambridge.   [2] Nicholson (1869), p. 497.

[3] Such Algae still exist in the genus *Crouania*, etc., in which morphologically though not structurally the habit is distinctly Sphenophyllaceous. The living *Lomentaria*, Gaill. is in habit very equisetaceous.

type of leaf which we associate more particularly with *Spheno-phyllum* is a much later development, unknown before the Upper Carboniferous. It appears to have arisen by a broadening of the narrow segments of the primitive type on exactly the same lines as we shall see were so frequently followed among Pteropsida.

We know very little of the fructifications of the earlier members of this race, but those of *Pseudobornia* (Fig. 26, 2, p. 54) appear to have possessed sporophylls essentially of a divided leafy type. At any rate in younger types, such as the Lower Carboniferous *Cheirostrobus* and the Upper Carboniferous *Sphenophyllum Dawsoni*, the sporophylls are divided, some lobes being fertile, others sterile. In the typical Sphenophyllaceous cone, one segment is sterile and protective and two segments (the sporangio-phores) are fertile. The sporangia, as in all those groups, are simply metamorphoses of parts of a fertile leaf-branch, just as they are in living ferns of to-day. Sometimes the whole segment is thus metamorphosed (when the sporangia are sessile), some-times only a part, while one portion remains as a sporangiophore.

As regards the stem, neither in the Sphenophyllales nor the Equisetales is there any trace of a pericaulome origin. The leaves being typically small in both these phyla, the stem appears to have acquired sufficient inherent mechanical strength without any such adaptation.

In the Equisetales, a group, which on the present evidence appears to be a little later in time than the Sphenophyllales, and unknown before the Lower Carboniferous, the leaves were primitively thalloid and forked. Those of *Archaeocalamites* come very near those of *Sphenophyllum* in this respect. The prevailing tendency in this phylum since then has been the reduction of such compound structures to a single segment (i.e. the leaves of *Calamites*), and at later periods to the almost complete union of these reduced leaf bases (e.g. *Equisetites* and *Equisetum*).

As regards the cones the tendency has been to sterilize certain leaf branches (the so-called bracts) alternating with fertile leaf branches. In *Archaeocalamites* these structures are absent, but in later cones, such as *Calamites*, they are always present and remain simple, undivided structures with the solitary exception of *Cingularia*.

The Evolution of the Pteropsida and Palaeophyllales.

The next group which we may consider is the Palaeophyllales, still entirely enigmatical. We may take these with the Pteropsida, because some Psygmophyllums so closely resemble some species of *Archaeopteris* that dispute has arisen as to the genus in which they should be included[1]. In *Psygmophyllum* (Fig. 27, p. 55) the leaf appears to have arisen by the flattening out of a large branch, in a wedge-shaped manner. The morphological unit here appears to be an axis bearing alternated metamorphosed branches, each branch being completely metamorphosed into a large cuneiform leaf, with the radiating nervation so common among primitive leaves. As a later development each leaf tends to become lobed or divided longitudinally.

Further the leaves are decurrent and thus the stem is to some extent pericaulomic.

Beyond this our knowledge of *Psygmophyllum* does not at present extend, and, in our ignorance of its fructification, it is for the moment maintained as the type of a distinct race.

Nearly all, if not all, known Pteropsida from Palaeozoic rocks are markedly pericaulomic and we agree with Potonié that in this group the stem is largely built up of leaf bases (cf. *Medullosa, Calamopitys,* and many other genera). Exact homologues of such stems are common among the Red Algae, e.g. *Polysiphonia.* It is clear also that this type of structure was common among members of the Psilophyton flora, for it is well seen in *Barrandeina* (Fig. 18, p. 36) and *Ptilophyton* (Fig. 12, p. 31) which appear to be Thallophytes.

The stem, in cases where large leaves are being evolved, would naturally require some accession to its mechanical strength, and this advantage would be gained by a pericaulome, a truly algal feature.

The differentiation of megaphylly from large branch structures appears to have taken place in several different directions. There

[1] This is the case with *Archaeopteris obtusa,* (Dawson), included in that genus by American Palaeobotanists (see Prosser (1894), Pl. II, p. 49) but transferred more recently by the present writer to *Psygmophyllum* (Arber (1912), p. 398) on the ground that in this type we appear to be dealing with leaves spirally arranged and not a pinnate type of frond.

is first the *Archaeopteris* type of leaf. This arose from the meta-morphosis of the branches of the *n*th order in a system which was at least bipinnate, if not more compound still. Each alternate branchlet was flattened out either into a sterile wedge-shaped leaf, with a radiating nervation, or metamorphosed into one or more sporangia. Every stage is clearly seen in Devonian species of *Archaeopteris* (Figs. 28–31, pp. 56–59). Further modifications of the wedge-shaped entire, primitive type of leaf soon set in. As in *Psygmophyllum*, these leaves tend to become lobed or segmented longitudinally.

In *A. Archetypus*, Schmalh. (Fig. 29, p. 57) and *A. Rogersi*, Daws., the leaflet is primi-tive, large and undivided. We next pass on to types such as *A. hibernica*, (Forbes) (Fig. 28, p. 56) and *A. Roemeriana*, (Goepp.) in which the leaf tends to become unsymmetrical and more or less toothed or lobed, and finally we reach such types as *A. fimbriata*, Nath. (Fig. 30, p. 58) and *A. fissilis*, Schmalh. (Fig. 31, p. 59) in which it is divided nearly to the base, longi-tudinally, into very narrow segments. A still further elaboration of this type by splitting would be indistinguishable from *Spheno-pteris*.

Further, other later genera, especially cha-racteristic of the Lower Carboniferous, such as *Rhacopteris*[1] (Figs. 42, and 33, p. 61), *Adiantites* (Fig. 43, p. 80) and *Cardiopteris*, have leaves essentially similar to *Archaeopteris* and with a radiating nervation. Further, here also the tendency to longitudinal splitting is marked at least in the two genera first named. Such types are in fact chiefly distinguished from *Archaeopteris* by the shape of the leaf and its segments—by some small peculiarity of its symmetry. They are all, however, obviously derived from the *Archaeopteris* type of leaflet.

Fig. 42. *Rhaco-pteris paniculi-fera*, Stur, from the Lower Car-boniferous of Austria. Fertile frond (reduced to ⅓ nat. size). After Stur, *Culm-Flora* (1875).

[1] In this genus the frond also dichotomises.

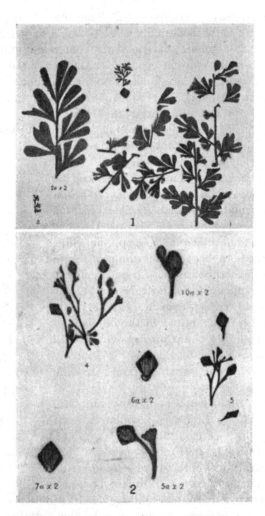

Fig. 43. *Adiantites fertilis*, (White), from the Lower Carboniferous of the United States. (1) Sterile fronds (reduced). (2) Fertile fronds and seeds (all natural size except where magnification is stated). After White, "Fossil Plants of the Group Cycado-filices," Smith's *Misc. Coll.*, Vol. 47, 1905.

Several attempts have been recently made to determine whether *Archaeopteris* was a member of the Primofilices or a Pteridosperm. Kidston[1] and others think the latter the more probable. We, however, agree with Johnson[2] that there is absolutely no evidence that this very completely known type (at any rate in the case of *A. hibernica*) bore seeds. It is much more probably a Primofilix. The evidence of *Rhacopteris*, so far as it goes, agrees with this conclusion. On the other hand at least one *Adiantites* (Fig. 43) was a seed-bearing plant, but here the seeds clearly arose by the metamor-phosis of a segment of the frond. No fructifications are yet known in the case of *Cardiopteris*.

Another very common type of frond evolution is seen in *Sphenopteridium* (Figs. 34, p. 61, 35, p. 62 and 44), beautifully fore-shadowed in the Psilophyton flora by *Pseudo-sporochnus* (Fig. 16, p. 34) and perhaps *Ptilo-phyton* (Fig. 13, p. 32), and fully developed in the *Archaeopteris* flora.

Here the ends of large branches frequently divide, usually dichotomously, to form a tuft of very narrow forked, nerved segments[3]. This type of frond persisted long after the Devonian period. It is very common in Lower Carboniferous rocks, e.g. *Sphenopteris affinis* (Fig. 45, p. 82) and *S. bifida* (Fig. 46, p. 82) among many others, and it is also met with in Upper Carboniferous times. The genus *Eremopteris* simply represents a somewhat peculiar modification of this type.

Fig. 44. *Sphenopterid-ium moravicum,*(Ett.), from the Lower Car-boniferous of Austria. Thalloid foliage (re-duced to ⅓ nat. size). After Stur, *Culm-Flora* (1875).

Between *Sphenopteridium* and *Sphenopteris* it is not possible to draw any really satisfactory line. The latter type possesses broader and more rounded segments as a rule, with a pinnate nervation. The term *Sphenopteridium* is in fact chiefly retained

---

[1] Kidston (1906), p. 435. [2] Johnson (1911[1]).
[3] A similar type of leaf also occurs where the division of an *Archaeopteris* or *Rhacopteris* type of pinnule is at its maximum.

because this particular type of leaf appears to be much more common than *Sphenopteris* in the older rocks. Sphenopterids however do occur there, and they appear to have been derived from Procormophytes by the division of the ends of flattened branch systems into more rounded and broader segments.

Fig. 45. *Sphenopteris affinis*, L. & H., from the Lower Carboniferous of Scotland, showing thalloid foliage. (Reduced ½ nat. size.) Specimen No. 667, Carboniferous Plant Coll., Sedgwick Museum, Cambridge. (W. Tams photo.)

Fig. 46. *Sphenopteris bifida*, L. & H., from the Lower Carboniferous of Scotland, showing thalloid foliage (× ⅘). Specimen V. 162, British Mus. (Nat. Hist.).

We know very little of the fructifications associated with either of these types of fronds in Devonian rocks. If Nathorst is right in attributing to Sphenopteridium-like fronds the tufted sporangia of *Cephalopteris* (Fig. 36, p. 63), then here at any rate

we have something different in the way of a fructification to
anything known among Carboniferous plants. On the other
hand, that of the Psilophytic genus *Bröggeria* (Fig. 17, p. 35)
has at least some superficial resemblance to that of *Cephalopteris*.
In both cases these fructifications may well arise from the
metamorphosis of a tuft of finely divided branchlets, and may

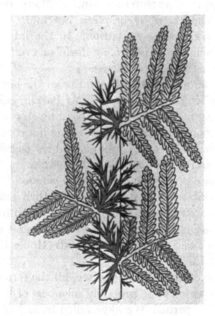

Fig. 47. Aphlebiae of *Pecopteris* (*Dacty-
lotheca*) *plumosa*, (Art.), from the Coal
Measures, showing persistence of thalloid
foliage. After Zeiller's *Élém. Paléobot.*
(1900).

be compared with those of *Archaeopteris* (Figs. 28, 2, p. 56 and
especially Fig. 32, p. 60).

It is thus clear that many of the earlier Pteropsida, like the
Sphenopsida, retained a sub-thalloid type of foliage. This type
further persisted among aphlebiae well into the Upper Carboniferous period. The aphlebiae of *Pecopteris* (Fig. 47) are a case
in point and are of the *Sphenopteridium* type, whereas many

other aphlebiae referred to the genus *Rhacophyllum* are extra-ordinarily algal in aspect. Even *Archaeopteris* itself retained not only scaly emergences at the base of the frond but a large pair of connate stipules (Fig. 28, 1, p. 56), no doubt originally of branch origin like the fused branches of *Cephalopteris*. The Archaeopteris type of leaflet also survived in the form of aphlebiae well into Upper Carboniferous times. Many of the smaller Cyclopterid pinnules of *Neuropteris* are more or less wedge-shaped with a radiating nervation. In the larger and ? older pinnules the shape changes to reniform or even orbicular, but the same type of nervation persists.

These facts afford another illustration of the law to which the present writer[1] in conjunction with Mr Parkin, called attention some years ago, namely that corresponding stages in the evolution of the various members of a plant are not contemporaneous in point of time. Here the foliar members clearly lag behind other organs and are of more primitive form.

It may be also pointed out that many genera of Carboniferous Pteropsida exhibit in the dichotomies of their fronds traces of their algal ancestry, as Potonié[2] long ago pointed out. Such genera as *Rhacopteris*, Fig. 33, p. 61 (Lower Carboniferous), and *Mariopteris* and many Sphenopterids (Upper Carboniferous) are cases in point.

Primofilices[3]. So far as we can see, all the fern-like plants of the Devonian period were probably members of the Primofilices and not Pteridosperms. We have failed to find any evidence of the latter group and the remarkable absence of fossils obviously of a seed nature has already been remarked upon here. However,

---

[1] Arber and Parkin (1907), p. 35.   [2] Potonié (1895).

[3] Prof. Seward (1910, p. 433) has taken exception to the term Primo-filices on the ground that, according to his view, the name implies "primary or primitive ferns." This is an entire misconception for which we think no justification will be found in the original paper by the writer (Arber, 1906) in which the term was first used. So far as we know, the adjective *primus* has never been interpreted as 'primitive' whereas it may imply 'early,' e.g. *prima nox*. The term Primofilices was and is intended simply to imply "early ferns," and also to suggest the "Primary or Palaeozoic Age of the race" (Arber (1906), p. 222). We fail to see any reason why this term should be abandoned for the name Coenopterideae which Seward adopts, especially as these plants do not appear to be generalised types.

until we know something of the structure of the fern-like plants of the Upper Devonian, it is impossible to feel confident as to their precise affinity. As things stand at present we have no certain evidence of the Pteridosperms before the Lower Carboniferous period. They certainly appear to have occurred then and in the case of the well-known *Adiantites fertilis,* (White) (Fig. 43, p. 80), it would appear that the seeds took their origin in much the same circumstances as the sporangia of the Primofilices, i.e. by the metamorphosis of part of a fertile branch, as in the Algae. No doubt the chief modifications in this case were the reduction of the number of megaspores in the sporangium to one and the development of an integument to the same. In *Adiantites 'ertilis* these features appear to have been arrived at very simply.

Cordaitales. There is one class of Pteropsida on the origin of which the Devonian floras as yet have thrown no light, namely the Cordaitales. A few obscure and fragmentary leaves, something like those of *Cordaites,* occur in Upper Devonian beds, but in the earlier floras we meet with nothing, so far as we can see, which appears to belong to this group. There is, it is true, the *Zosterophyllum* of the Lower Devonian (Fig. 22, p. 41), with long parallel-sided leaves or axes, but the fossil is so obscure at present that no conclusion of any value can be drawn from it. Even from Lower Carboniferous rocks evidence of the Cordaitales is at present extremely scanty, and until we learn more about the group at this period, we are not likely to trace it back to Devonian times.

The Evolution of the Lycopsida.

The evolution of the Lycopsida along the lines first perceived by Lignier, is particularly well seen among members of the Psilophyton flora. From such a type as *Psilophyton* (Figs. 3, p. 17, and 5, p. 19) with its chaffy scattered scales, it is but a short step to *Thursophyton* (Fig. 10, p. 28), which is to all intents and purposes a Lycopod. Even if this type was not vascular, either in axis or leaf, it is quite easy now, in the light of *Psilophyton,* to understand how similar vascular types may have arisen. From *Thursophyton* it is a short step to *Protolepidodendron.*

In this group a pericaulome was not always evolved. *Bothro-dendron*, one of the earliest Devonian Lycopods (Fig. 37, p. 64), does not possess this feature, which however occurs in *Archaeo-sigillaria* (Fig. 38, p. 66) and *Leptophloeum* (Fig. 40, p. 67) of the same period. Most of the later Palaeozoic Lycopods, such as *Lepidodendron*, as well as many Sigillarias, possess a pericaulome, but we see how in the later types of the latter genus (section *Clathraria*) the pericaulome was abandoned.

In the Lycopsida the leaves, arising from emergences, are always unbranched and independent. The sporangia were originally borne at the base of the upper (adaxial) surface of the leaf, as we see them in *Thursophyton* (Fig. 11, p. 30) and as indeed has been the rule among Lycopods ever since. Among Lycopods the *Selago* condition is thus primitive as regards the place of origin of the sporangia.

We are able to throw no light on the original morphology and function of the ligule. Such an organ is universally present in Palaeozoic Lycopods, but it is there always in much the same condition as in the living *Selaginella*.

The pseudodichotomous branching of many Palaeozoic Lycopods is another indication of algal ancestry.

The Psilotales.

There remains for consideration one further group, the Psilotales, which may be dealt with here, since by some authorities these plants are regarded as more nearly related to the Lycopsida than to any other group. There are also further grounds for a discussion of this race arising from the fact that Kidston and Lang[1] have recently instituted a comparison between *Psilophyton* and members of the Psilotales, and have drawn the conclusion that some phylogenetic connexion may exist between them.

On the much debated question of the general affinity of the Psilotales, we agree with Mrs Thoday[2] and others that they are best regarded as an entirely distinct group from either the Lycopods or the Sphenophylls. We should not include them in either the Lycopsida or Sphenopsida. At the same time we do

---

[1] Kidston and Lang (1917), p. 776.   [2] Sykes (1908).

not deny that there are points of similarity between the Psilotales
and both those groups. These resemblances however appear to
us to be far outweighed by the differences, and the latter seem
to be of a more fundamental nature than the former.

In comparison with the Lycopsida, we would point to the
microphyllous habit and the presence of a pericaulome, as
features in common. On the other hand in the anatomy of their
sporophyte and its fructifications the Psilotales are quite unlike
living or fossil Lycopods.

In recent years the view that this group may be related to
the Sphenopsida rather than to the Lycopsida has undoubtedly
gained ground, despite the vast differences in habit. In certain
points in the anatomy of the shoot and in the type of sporophyll
it is possible to institute some comparison between the Psilotales
and the Sphenophyllales[1]. We are now inclined to regard these,
however, as cases of parallelism of development and not to
attach any special phyletic arguments to such resemblances
which are far outweighed by more important dissimilarities[2].
We should be inclined to regard the Psilotales as a quite independent race, also of algal origin, which appeared on the scene
long after the other races which we are discussing here, possibly
in Mesozoic times or even later. We know of no evidence (not-
withstanding Kidston and Lang's views of *Psilotum* to which
we shall shortly refer again) of the existence of any plant in
Palaeozoic times which has any real claim to inclusion in this
group. Certainly none occurs among the known members of the
Psilophyton or Archaeopteris floras. The Psilotales, like the
Bryophyta, are thus a much later group in point of age than the
Sphenopsida, Lycopsida and Pteropsida. That the Psilotales
retain certain features which may be primitive we should agree,
though certainly in *Psilotum* and probably also in *Tmesipteris*
other characters, which might be regarded at first sight as

---

[1] Scott (1909), Vol. ii, p. 626.

[2] The most recent advances of our knowledge of the Psilotales which we
owe to Lawson (1917) do not help us here since they are concerned with
the gametophytes. The latter do not appear to be very remote from those
of Lycopsida, but since the corresponding structures of the Sphenophyllales
are not known and probably never will be known, we have no grounds for
any comparison in this respect.

primitive, are, in our opinion, reduction features, the reduction being correlated with the semi-saprophytic habit of these plants.

As primitive features we would point to the microphyllous habit and pericaulome, characters which as we have already seen, appear to have arisen in different groups quite independently. The forking of the sporophylls is another such feature, originating as in other groups from dichotomous thalloid branches. In this feature the Psilotales may resemble Sphenophylls rather than Lycopods as has often been pointed out, and it may be that in this group the leaf was originally a branch and not an emergence as in the Lycopods. But even if this is the case, it is more than likely, in our view, that these features were quite independent in origin from the similar organs of Sphenophylls, especially since the latter race was in all probability entirely defunct long before the Psilotales appeared on the scene.

The vascular structure of the Psilotales represents a primitive stage passed through in the history of the evolution of the stele in many, if not all groups, and here either not developed further or regained by reduction in correlation with the semi-saprophytic habit.

Finally as regards *Psilophyton* we see no real point of contact with the Psilotales except in habit, a character which of course is worse than useless as a guide to affinity, if considered by itself. In the first place the fructification and the manner in which it is borne are entirely distinct. *Tmesipteris* is a Cormophyte without emergences, *Psilophyton* is a Thallophyte possessing emergences, and *Psilotum* is in all probability a reduced semi-leafless Cormophyte. Finally the only comparison as regards the stele of *Psilophyton* is confined to the tip of the shoot in the Psilotales, near the growing point. In the mature stems of the latter the stele has well-developed primary wood, arising from several protoxylem groups (unknown in *Psilophyton*) and has altogether reached a higher stage of development than anything known in *Psilophyton*. At the same time, as we shall show in the following chapter, this being a stage passed through in the evolution of the steles of most groups, has no obvious phyletic bearing.

CHAPTER VII

THE ORIGIN OF THE STELE IN THE EARLIER CORMOPHYTA

THE discovery, or rather the confirmation of the discovery, of an extremely primitive stele in species of *Psilophyton* and also of *Arthrostigma* has naturally a very important bearing on our notions of the origin of the stele in Pteridophyta.

Among the oldest fossil plants, of far greater antiquity than the very ancient land plants under consideration here, there are many multi-cellular types without any trace of a specialised conducting tissue. These fossils commonly occur in marine rocks and they so closely resemble types of living Algae in their general structure that we are justified in regarding them as marine Algae.

Somewhat later in point of time, in the marine Silurian rocks and extending into the early Devonian, we find another type of anatomical habit in which the whole thallus is tubular. This type is represented in *Nematophycus* where the tubes are sometimes of two different calibres, the smaller forming a dense packing between the larger tubes. This is however a feature also common to some living Algae and needs no further remark here.

Among living Algae no more advanced type of water conducting tissue is met with, though some Algae have evolved what we regard as a quite typical phloem. The fact that nearly all the higher living Algae, other than the symbiotic types (Lichens) are hydrophytes and not terrestrial plants, furnishes a ready explanation of the absence of a lignified conducting tissue. There is no need of such a tissue.

When however we come to Devonian land plants we find in *Psilophyton* and *Arthrostigma* and other types an extremely early stage in the evolution of the stele. What we have is a single protoxylem group alone, formed by the simultaneous modification of a set of procambial elements. That modification was not continuous or progressive. A certain small set of procambial elements were converted together and then the process ended. There is not the slightest sign of a leaf trace origin of the cauline stele. In fact we know from *Psilophyton* that the conversion of the procambial elements took place in the main axis and in the branches quite independently (probably at different periods) and further that these foci of lignification were not in continuity except by means of parenchymatous tissues. Thus here at any rate the axes existed first before their steles. Continuity between the steles of members of the different orders of axes was a later modification, the usefulness of which soon became obvious.

We may picture *Psilophyton* as a terrestrial plant growing under damp conditions in a bed of peat, and no doubt in such a habitat a partial, little-developed and discontinuous conducting system met the case.

The next stage in the evolution of the stele appears to have been the substitution of a continuous for a purely initial transformation of procambial elements. When the first set of elements had been converted (protoxylem) the process continued and primary wood, focused on the single protoxylem, was evolved. This stage is thus the evolution of a monarch strand. Such a stage exists for instance in the rootlet (*not* the rhizophore) of Stigmaria.

The next step appears to have been different in different cases. The procambial activity having ceased, a secondary cambium was sometimes (but probably rarely) initiated and secondary wood was added to the framework of a single monarch strand. This state of affairs is also perceived in certain Stigmarian rootlets.

A more common and perhaps more successful course of stelar evolution in stems was the substitution of a number of protoxylems with prolonged cambial activity in the central region of the stem in place of a single such group. These groups may have been centric or excentric, in the latter case the groups being concentric.

In the case of centric groups of protoxylem, space was naturally limited, and even where the groups were excentric but concentric, the room for the development of primary tissues was limited. In either case a stage was eventually reached when further vascular development could only be provided for by peripheral additions, i.e. secondary wood. This new type of structure, as is well known, was often accompanied by a reduction in the primary scaffolding.

In drawing attention to the existence in Devonian times of a primitive type of stele, it must not be imagined that all Devonian plants possessed such anatomy. On the contrary we know of several arborescent types possessing secondary wood of the modern type found among the higher plants, occurring in the Middle Devonian of Scotland[1] and the Devonian of Russia[2] and the United States[3] and Canada[4]. It is thus clear that alongside of *Psilophyton* there existed other plants which had already reached a far higher stage of evolution as regards the stele than we meet with in that genus. In fact *Psilophyton* was probably one of the latest survivals of the primitive Devonian types[5].

[1] *Palaeopitys Millerii*, M'Nab (1870).
[2] *Callixylon Trifilievi* (= *Dadoxylon Trifilievi*), Zalessky (1909) and (1911), p. 28, Pl. IV, figs. 1–3.
[3] *Callixylon Oweni*, Elkins and Wieland (1914).
[4] Dawson (1871).
[5] [The Author left this chapter unfinished, and regarded it merely as a preliminary draft. A.A.]

BIBLIOGRAPHY

ARBER, E. A. N. (1906). On the Past History of the Ferns. Ann. Bot. Vol. xx. p. 215. 1906.

—— (1912). On *Psygmophyllum majus*, sp. nov. from the Lower Carboniferous rocks of Newfoundland, together with a Revision of the Genus and Remarks on its Affinities. Trans. Linn. Soc. London, Bot. Ser. 2, Vol. VII. Pt 18, p. 391. 1912.

ARBER, E. A. N. and GOODE, R. H. (1915). On some Fossil Plants from the Devonian Rocks of North Devon. Proc. Camb. Phil. Soc. Vol. XVIII. Pt 3, p. 89. 1915.

ARBER, E. A. N. and PARKIN, J. (1907). On the Origin of Angiosperms. Journ. Linn. Soc. London, Bot. Vol. 38, p. 29. 1907.

CARRUTHERS, W. (1872). Notes on some Fossil Plants. Geol. Mag. Dec. 1, Vol. 9, p. 49. 1872.

—— (1873). On some Lycopodiaceous Plants from the Old Red Sandstone of the North of Scotland. Journ. Bot. Vol. XI. (N.S. Vol. 2), p. 321. 1873.

CRAMPTON, C. B. and CARRUTHERS, R. G. (1914). The Geology of Caithness. Mem. Geol. Surv. Scotland, Nos. 110, 116. 1914.

CRÉPIN, F. (1874). Description de quelques plantes fossiles de l'étage des psammites du Condroz (dévonien supérieur). Bull. Acad. R. Belgique, Ser. 2, Vol. 38, p. 356. 1874.

—— (1875). Observations sur quelques plantes fossiles des dépôts dévoniens, etc. Bull. Soc. roy. Bot. Belgique, Vol. XIV. p. 214. 1875.

DAVID, T. W. E. and PITTMAN, E. F. (1893). On the Occurrence of Lepidodendron Australe? in the Devonian rocks of New South Wales. Rec. Geol. Surv. N.S. Wales, Vol. III. N.S. Pt 4, p. 194. 1893.

DAWSON, J. W. (1859). On Fossil Plants from the Devonian Rocks of Canada. Quart. Journ. Geol. Soc. Vol. 15, p. 477. 1859.

—— (1871). The Fossil Plants of the Devonian and Upper Silurian Formations of Canada. Part I. Geol. Surv. Canada. 1871.

—— (1878). Notes on some Scottish Devonian Plants. Canadian Natural. N.S. Vol. VIII. p. 379. 1878.

—— (1888). The Geological History of Plants. (Intern. Sci. Series), London. 1888.

DAWSON, SIR W. and PENHALLOW, D. P. (1891). *Parka decipiens*. Notes on specimens from the collections of James Reid, Esq. Trans. Roy. Soc. Canada, Vol. IX. Sect. IV. p. 3. 1891.

DON, A. W. R. and HICKLING, G. (1917). On *Parka decipiens*. Quart. Journ. Geol. Soc. Vol. 71, p. 648. 1917.

DUN, W. S. (1897). On the Occurrence of Devonian Plant-bearing beds on the Genoa River, County of Auckland. Rec. Geol. Surv. N.S. Wales, Vol. V. Pt 3, p. 113. 1897.

ELKINS, M. G. and WIELAND, G. R. (1914). Cordaitean Wood from the Indiana Black Shale. Amer. Journ. Sci. Vol. 38, p. 65. 1914.

FLEMING, J. (1831). On the occurrence of the Scales of Vertebrated Animals in the Old Red Sandstone of Fifeshire. Cheek's Edinburgh Journ. Nat. and Geogr. Sci., N.S. Vol. III. p. 86. 1831.

FRECH, F. (1897). Lethaea geognostica. Theil I. Lethaea Palaeozoica. Vol. II. 1897–1902.

FRITSCH, F. E. (1916). The Algal Ancestry of the Higher Plants. New Phytol. Vol. XV. p. 233. 1916.

GILKINET, A. (1875¹). Sur quelques plantes fossiles de l'étage des psammites du Condroz. Bull. Acad. R. Belgique, Ser. 2, Vol. 39, p. 384. 1875.

—— (1875²). Sur quelques plantes fossiles de l'étage du Poudingue de Burnot (dévonien inférieur). Bull. Acad. R. Belgique, Ser. 2, Vol. 40, p. 139. 1875.

HALLE, T. G. (1916). Lower Devonian Plants from Röragen in Norway. Handl. k. Svenska Vetenskaps-Akad. Vol. 57, No. 1. 1916.

HICKLING, G. (1908). The Old Red Sandstone of Forfarshire, Upper and Lower. Geol. Mag. Dec. 5, Vol. V. p. 396. 1908.

—— (1912¹). The Old Red Sandstone Rocks near Arbroath. Proc. Geol. Assoc. Vol. 23, p. 299. 1912.

—— (1912²). On the Geology and Palaeontology of Forfarshire. Proc. Geol. Assoc. Vol. XXIII. p. 302. 1912.

HINXMAN, L. W. and GRANT WILSON, J. S. (1902). The Geology of Lower Strathspey. Mem. Geol. Surv. Scotland. 1902.

HORNE, J. and HINXMAN, L. W. (1914). The Geology of the Country round Beauly and Inverness. Mem. Geol. Surv. Scotland, No. 83. 1914.

HORNE, J. and MACKIE, W. (1917). The Plant-bearing Cherts at Rhynie. Aberdeenshire. Report of the Committee. Rep. Brit. Assoc. Newcastle-on-Tyne, 1916, p. 206. 1917.

JAHN, J. J. (1903). Ueber die Etage H im mittel-böhmischen Devon. Verhandl. K. K. Geol. Reichsanstalt, Jahrg. 1903, p. 73. 1903.

JANCHEN, E. (1911). Neuere Vorstellungen über die Phylogenie der Pteridophyton. Mitt. des Naturwiss. Ver. Univ. Wien, Jahrg. IX. p. 33. 1911.

JOHNSON, T. (1911¹). Is Archaeopteris a Pteridosperm? Sci. Proc. R. Dublin Soc. Vol. XIII. (N.S.), No. 8, p. 114. 1911.

—— (1911²). The Occurrence of Archaeopteris Tschermaki, Stur, and of other species of Archaeopteris in Ireland. Sci. Proc. R. Dublin Soc. Vol. XIII. (N.S.), No. 9, p. 137. 1911.

—— (1913). On Bothrodendron (Cyclostigma) Kiltorkense, Haughton, sp. Sci. Proc. R. Dublin Soc. Vol. XIII. (N.S.), No. 34, p. 500. 1913.

KIDSTON, R. (1885). On the Occurrence of Lycopodites (Sigillaria) Vanuxemi, Göppert, in Britain with remarks on its Affinities. Journ. Linn. Soc. Bot. Vol. 21, p. 560. 1885.

—— (1888). On the Fructification and Affinities of Archaeopteris hibernica, Forbes, sp. Ann. Mag. Nat. Hist. Ser. 6, Vol. I. p. 412. 1888.

—— (1893). On the Occurrence of Arthrostigma gracile, Dawson, in the Lower Old Red Sandstone of Perthshire. Proc. R. Phys. Soc. Edinburgh, Vol. XII. p. 102. 1894 (for 1893.)

94   BIBLIOGRAPHY

KIDSTON, R. (1897). On *Cryptoxylon Forfarense*, a New Species of Fossil Plant from the Old Red Sandstone. Proc. Roy. Phys. Soc. Edinb. Vol. XIII. p. 360. 1897.
—— (1902). Note on the Fossil Plants of the Old Red Sandstone of Scotland, in Hinxman, L. W. and Grant Wilson, J. S. The Geology of Lower Strathspey, p. 83. Mem. Geol. Surv. Scotland. 1902.
—— (1906). On the Microsporangia of the Pteridospermeae, with Remarks on their Relationship to Existing Groups. Phil. Trans. Roy. Soc. Ser. B, Vol. 198, p. 413. 1906.
KIDSTON, R. and LANG, W. H. (1917). On Old Red Sandstone Plants showing Structure, from the Rhynie Chert Bed, Aberdeenshire. Part I. Rhynia Gwynne-Vaughani, Kidston and Lang. Trans. Roy. Soc. Edinb. Vol. 51, Pt 3, p. 761. 1917.
KREJČI, J. (1879). Notiz über die Reste von Landpflanzen in der Böhmischen Silurformation. Sitzungsber. k. Böhm. Gesell. Wissen. Prag, Jahrg. 1879, p. 201. 1880.
KUBART, B. (1913). Zur Frage der Perikaulomtheorie. Ber. Deutsch. Bot. Gesell. Vol. 31, p. 567. 1913.
LAWSON, A. A. (1917). The Gametophyte Generation of the Psilotaceae. Trans. Roy. Soc. Edinb. Vol. 52, Pt 1, p. 93. 1917.
LIGNIER, O. (1903). Equisétales et Sphénophyllales. Leur origine filicinéenne commune. Bull. Soc. Linn. Normandie, Ser. 5, Vol. 7, p. 93. 1903.
—— (1908). Sur l'origine des Sphénophyllées. Bull. Soc. Bot. France, Vol. 55, p. 278. 1908.
—— (1909). Essai sur l'Évolution morphologique du Règne végétal. Compte Rendu Ass. franç. Avanc. Sci. 37ᵉ Session, Clermont-Ferrand, 1908, p. 530. 1909. See also Bull. Soc. Linn. Normandie, Ser. 6, Vol. 3, Pt 2, p. 35. 1909.
LUDWIG, R. (1869). Fossile Pflanzenreste aus den paläolithischen Formationen der Umgegend von Dillenburg, Biedenkopf und Friedberg und aus dem Saalfeldischen. Palaeontogr. Vol. 17, No. 3, p. 105. 1869.
MACKIE, W. (1914). The Rock Series of Craigbeg and Ord Hill, Rhynie, Aberdeenshire. Trans. Edinb. Geol. Soc. Vol. x. p. 205. 1914.
M'NAB, W. R. (1870). On the Structure of a Lignite from the Old Red Sandstone. Trans. Bot. Soc. Edinburgh, Vol. x. p. 312. 1870.
MACNAIR, P. and REID, J. (1896). On the Physical Conditions under which the Old Red Sandstone of Scotland was deposited. Geol. Mag. Dec. 4, Vol. III. p. 106. 1896.
MILLER, HUGH (1841). The Old Red Sandstone. 1st Edit. Edinburgh, 1841.
—— (1857). The Testimony of the Rocks. 1st Edit. Edinburgh, 1857.
NATHORST, A. G. (1894). Zur Paläozoischen Flora der Arktischen Zone. Handl. k. Svenska Vetenskaps-Akad. Vol. 26, No. 4. 1894.
—— (1902). Zur Oberdevonischen Flora der Bären-Insel. Handl. k. Svenska Vetenskaps-Akad. Vol. 36, No. 3. 1902.
—— (1904). Die Oberdevonische Flora des Ellesmerelandes. Rep. 2nd Norwegian Arctic Exped. *Fram*, 1898–1902, No. 1. 1904.
—— (1913). Die Pflanzenreste der Röragen-ablagerung, in Goldschmidt, V. M., Das Devongebiet am Röragen bei Röros. Videnskap. Skrift. Kristiania, Vol. I. Mat.-Nat. Klasse, No. 9. 1913.

NATHORST, A. G. (1915). Zur Devonflora des westlichen Norwegens. Bergens Mus. Aarbok, 1914–5, No. 9. 1915.

NEWBERRY, J. S. (1889). Devonian Plants from Ohio. Journ. Cincinnati Soc. Nat. Hist. Vol. XII. p. 48. 1889.

NICHOLSON, H. A. (1869). On the Occurrence of Plants in the Skiddaw Slates. Geol. Mag. Dec. 1, Vol. VI. p. 494. 1869.

NICHOLSON, H. A. and LYDEKKER, R. (1889). Manual of Palaeontology. Edit. 3, Vol. 2. 1889.

PENHALLOW, D. P. (1892). Additional Notes on Devonian Plants from Scotland. Canadian Rec. Sci. Vol. 5, No. 1, pp. 1–13, 2 pls. 1892.

POTONIÉ, H. (1895). Die Beziehung zwischen dem echt-gabeligen und dem fiederigen Wedel-Aufbau der Farne. Ber. Deutsch. Botan. Gesell. Vol. XIII. p. 244. 1895.

—— (1898). Die Metamorphose der Pflanzen im Lichte palaeontologischer Thatsachen. Berlin. 1898.

—— (1901). Die Silur- und die Culm-Flora des Harzes und des Magdeburgischen. Abhandl. k. Preuss. Geol. Landesanstalt, N.F. Heft 36. 1901.

—— (1902[1]). Die Pericaulom-Theorie. Ber. Deutsch. Botan. Gesell. Vol. 20, p. 502. 1902.

—— (1902[2]). Ein Blick im die Geschichte der botanischen Morphologie mit besonderer Rücksicht auf die Pericaulom-Theorie. Naturwiss. Wochenschrift, Vol. II. (N.S.), Nos. 1–3. 1902.

POTONIÉ, H. and BERNARD, C. (1904). Flore Dévonienne de l'étage H de Barrande. Leipzig, 1904. (Syst. Silur. du Centre de la Bohème, by J. Barrande.)

PROSSER, C. S. (1894). The Devonian System of Eastern Pennsylvania and New York. Bull. No. 120, U.S. Geol. Survey. 1894.

REID, J., GRAHAM, W. and MACNAIR, P. (1897). *Parka decipiens*, its Origin, Affinities and Distribution. Trans. Geol. Soc. Glasgow, Vol. XI. Pt 1, p. 105. 1897.

REID, J. and MACNAIR, P. (1896). On the Genera *Lycopodites* and *Psilophyton* of the Old Red Sandstone Formation of Scotland. Trans. Geol. Soc. Glasgow, Vol. X. Pt 2, p. 323. 1896.

—— (1899). On the genera Psilophyton, Lycopodites, Zosterophyllum and Parka decipiens of the Old Red Sandstone of Scotland. Their Affinities and Distribution. Trans. Edinb. Geol. Soc. Vol. VII. p. 368. 1899 (for 1893–1898).

SALTER, J. W. (1858). On some Remains of Terrestrial Plants in the Old Red Sandstone of Caithness. Quart. Journ. Geol. Soc. Vol. 14, p. 72. 1858.

SALTER, J. W. in MURCHISON, R. I. (1859). On the Succession of the Older Rocks in the Northernmost Counties of Scotland, etc. Quart. Journ. Geol. Soc. Vol. 15, p. 407. 1859.

SCHMALHAUSEN, J. (1894). Ueber Devonische Pflanzen aus dem Donetz-Becken. Mém. Com. Géol. St-Pétersbourg, Vol. VIII. No. 3, 36 pp. and 2 pls. 1894.

SCOTT, D. H. (1909). Studies in Fossil Botany. Edit. 2, Vol. II. London. 1909.

SCOTT, D. H. (1910). Presidential Address. Proc. Linn. Soc. London, 123rd Session, p. 17. 1911 (for 1910–11).

SEWARD, A. C. (1910). Fossil Plants. Vol. II. Cambridge, 1910.

SMITH, G. O. and WHITE, D. (1905). The Geology of the Perry Basin in South-eastern Maine. Prof. Paper, No. 35, U.S. Geol. Surv. 1905.

SOLMS LAUBACH, GRAF ZU (1895). Ueber devonische Pflanzenreste aus den Lennschiefern der Gegend von Gräfrath am Niederrhein. Jahrb. K. Preuss. Geol. Landesanst. für 1894 (Abh. v. ausserhalb d. k. Geol. Landes, etc.), p. 67. 1895.

STUR, D. (1881). Die Silur- Flora der Etage H–h₁ in Böhmen. Sitzungsber. K. Akad. Wissen. Vol. 84, Pt 1, p. 330. 1881.

SYKES, M. G. (Mrs Thoday) (1908). The Anatomy and Morphology of Tmesipteris. Ann. Bot. Vol. 22, p. 63. 1908.

WHITE, DAVID (1905). See SMITH, G. O. and WHITE, D. (1905).

—— (1907). A remarkable Fossil Tree Trunk from the Middle Devonic of New York. Bull. New York State Museum Geol. Papers, No. 107, p. 327. 1907.

ZALESSKY, M. D. (1909). Communication préliminaire sur un nouveau *Dadoxylon* à faisceaux de bois primaire autour de la moelle, provenant de dévonien supérieur du bassin du Donetz. Bull. Acad. Impér. Sci. St-Pétersbourg, Ser. VI. Vol. III. p. 1175. 1909.

—— (1911). Étude sur l'anatomie du *Dadoxylon Tchihatcheffi*, Goeppert, sp. Mém. du Comité Géol. St-Pétersbourg, Nouvelle Série, Livraison 68. 1911.

INDEX

The principal references are in black type

Aberdeenshire, 5
Adiantites, 69, 79, 80 (Fig. 43), 81
Adiantites fertilis, 80 (Fig. 43), 85
Africa, South, 58
Age, geological, of Devonian Floras, 3–7
Algae, 39, 40, 42, 47, 49, 70, 71, 73, 74, 76, 78, 84, 89
America, 17, 37, 66, 68, 69
Annularia, 75, 76
Aphlebiae, 83 (Fig. 47), 84
Aphyllopteris, 12, **43, 44**
Archaeocalamites, 69, 77
Archaeopteris, 9, 11, 12, 13, **56** (Fig. 28), **57** (Fig. 29), **58** (Fig. 30), **59** (Fig. 31), **60** (Fig. 32), 67, 68, 79, 81, 83, 84
Archaeopteris Flora, 1, **9–10**, 11, 12, 51, **52–68**, 87
Archaeopteris Archetypus, 12, 57 (Fig. 29), 79
A. fimbriata, 58 (Fig. 30), 79
A. fissilis, 12, 59 (Fig. 31), 79
A. hibernica, 10, 11, 56 (Fig. 28), 79, 81
A. Hitchcocki, 60 (Fig. 32)
A. Howitti, 13
A. Jacksoni, 13
A. Roemeriana, 12, 79
A. Rogersi, 13, 79
A. Tschermaki, 11
A. Wilkinsoni, 13
Archaeosigillaria, 9, 10, 12, 13, **65**, **66** (Figs. **38, 39**), 67, 68, 69, 86
Archaeosigillaria primaeva, 66 (Fig. 39)
A. Vanuxemi, 66 (Fig. 38)
Arctic Regions, 12, 68
Arenig, 76
Arthrostigma, 8, 10, 11, 12, 13, **26, 27** (Figs. **8 and 9**), 28 (Fig. 10), 29, 47, 50, 51, 72, 74, 89
Arthrostigma gracile, 11, 27 (Figs. 8 and 9), 46
Asterolepis, 4
Auckland, 13
Australia, 6, 7, 13, 38, 67, 68, 69

Baggy Beds, 5
Barinophyton, 9, 11, 13, **37, 38** (Fig. **19**), 68

Barinophyton Richardsoni, 38 (Fig. 19)
Barrande, J., 5, 12
Barrandeina, 9, 10, 12, 13, **36** (Fig. **18**), **37**, 78
Barrandeina Dusliana, 36 (Fig. 18)
Batrachospermum, 74
Bear Island, 2, 6, 7, 12, 43, 52, 62, 64
Belgium, 5, 7, 11, 16, 26, 38, 64
Bernard, C., 30, 35, 50
Bohemia, 2, 3, 5, 7, 12, 26, 29, 30, 35, 37, 42, 43
Bothrodendron, 9, 13, **64** (Fig. **37**), 65, 86
Bothrodendron Kiltorkense, 11, 12, **64** (Fig. **37**), 65
Bower, F. O., 70
Bröggeria, 10, 12, **35** (Fig. **17**), 37, 50, 83
Bröggeria norvegica, 35 (Fig. 17)
Brongniart, A., 48, 64
Bryophyta, 44, 70, 71, 72, 87
Buthotrephis radiata, 76

Caithness, 10
Calamites, 65, 69, 72, 77
Calamopitys, 78
Canada, 6, 7, 8, 13, 14, 16, 26, 38, 56, 67, 68, 75, 91
Carboniferous, Lower, 9, 14, 52, 56, 66, 67, **69**, 77, 81, 84, 85
Carboniferous, Upper, 52, 56, 62, 65, 77, 81, 83, 84
Cardiopteris, 69, 79, 81
Carpolithes, 68
Carruthers, W., 17
Caulopteris, 13, 37
Caulopteris Peachii, 10
Cephalaspis, 4
Cephalopteris, 9, **62, 63** (Fig. **36**), 64, 83, 84
Cephalopteris mirabilis, 12, 63 (Fig. 36)
Cheirolepis, 4
Cheirostrobus, 77
Chemung series, 6, 13
China, 6
Cingularia, 77
Clathraria, 86
Coal Measures, 9, 58, 60, 61, 64, 68
Coccosteus, 4

Printed in the United States
By Bookmasters